Dennis DiClaudio is the au~~thor~~ ~~chondriac's~~
*Pocket Guide to Horrible Diseases You Probably Already
Have, The Paranoid's Pocket Guide to Mental Disorders
You Can Just Feel Coming On* and *The Deviant's Pocket
Guide to the Outlandish Sexual Desires Barely Con-
tained in Your Subconscious.* He's currently a writer for
Indecision2008.com, as well as several other Comedy
Central Web sites, and an improvisational comedian. He
lives in New York City, though his mind is often in Phil-
adelphia wandering around somewhere near the Italian
Market.

MAN
VS.
WEATHER

How to Be Your Own Weatherman

Dennis DiClaudio

PENGUIN BOOKS

For my older brother,

Jason Toogood

PENGUIN BOOKS
Published by the Penguin Group
Penguin Group (USA) Inc., 375 Hudson Street,
New York, New York 10014, U.S.A.
Penguin Group (Canada), 90 Eglinton Avenue East, Suite 700, Toronto,
Ontario, Canada M4P 2Y3 (a division of Pearson Penguin Canada Inc.)
Penguin Books Ltd, 80 Strand, London WC2R 0RL, England
Penguin Ireland, 25 St Stephen's Green, Dublin 2, Ireland
(a division of Penguin Books Ltd)
Penguin Group (Australia), 250 Camberwell Road, Camberwell,
Victoria 3124, Australia (a division of Pearson Australia Group Pty Ltd)
Penguin Books India Pvt Ltd, 11 Community Centre, Panchsheel Park,
New Delhi – 110 017, India
Penguin Group (NZ), 67 Apollo Drive, Rosedale, North Shore 0632,
New Zealand (a division of Pearson New Zealand Ltd)
Penguin Books (South Africa) (Pty) Ltd, 24 Sturdee Avenue, Rosebank,
Johannesburg 2196, South Africa

Penguin Books Ltd, Registered Offices:
80 Strand, London WC2R 0RL, England

First published in Penguin Books 2008

10 9 8 7 6 5 4 3 2 1

Copyright © Dennis DiClaudio, 2008
All rights reserved

Image credits
Pages 148 and 149: iStockphoto.com
Chart on pages 194–195: From *Meteorology Today*, seventh edition, by
C. Donald Ahrens. © 2003 Brooks/Cole, a part of Cengage Learning, Inc.
Reproduced by permission. www.cengage.com/permissions
Pages 194 and 195: the author
All other illustrations: Michael Alexander Moser

ISBN 978-0-14-311363-8
CIP data available

Printed in the United States of America
Set in ITC Bookman Light along with Goudy Sans
Designed by Sabrina Bowers

CONTENTS

INTRODUCTION
THE INTRODUCTION

On August 23, 2005, two dangerously low-pressure systems—a tropical wave and the remains of a dying tropical depression—met up with each other somewhere over The Islands of the Bahamas in the eastern Atlantic Ocean. The two systems fused themselves to each other, combining energy to form Tropical Storm Katrina. Sucking heat up from the ocean as it rolled toward the Gulf of Mexico and the southeastern United States, it grew stronger, fiercer, blowier. Just two hours before it made landfall in Florida on August 25, it had managed to collect enough energy to build itself into a Category 1 hurricane.

It ripped through Miami-Dade County with 80 mph winds, pelting Floridians with fourteen inches of rain, tearing down homes, uprooting trees, and unleashing a plague of politicians in its wake. Florida got off easy.

The hurricane skipped back out into the Gulf to recharge itself—the effects of its winds continuing to cause damage to cities along the Gulf Coast and northern Cuba—only to hit land once again four days later in Louisiana. But this time it was a Category 3 hurricane.

Spiraling 125 mph winds surged through the low-lying spicy Cajun landscape, swamping it in ten inches of rainfall. Electricity went out, windows were blown shattering from skyscrapers, and waves crashed over the feckless levees designed to keep the Gulf waters at bay. And then those very levees broke.

Water burst into New Orleans like a wet invading army, flooding entire neighborhoods, destroying homes, and sending residents onto their roofs for sanctuary. The greater part of the city was evacuated.

Ultimately Hurricane Katrina took more than seventeen hundred human lives across several states. The tendrils of its destruction reached as far away as Huron County, Ohio. Damage estimates are calculated in the billions of dollars. And New Orleans—the Big Easy, the birthplace of jazz, one of the top 170 finest cities in the country—was practically decimated.

As you can plainly see, Weather is a total douche bag.

Somebody has to do something to stop it.

Luckily I just did.

CHAPTER 1
KNOCK KNOCK—
WHO'S THERE?—
IT'S WEATHER!

Make no mistake about it: Weather is *out there*. And it is waiting for you.

No doubt you feel safe now, reclining in your living room with this nifty little book about weather you recently acquired. Oh, it'll most likely be good for a few facts about barometric pressure and the stratosphere and which direction rain falls. Stuff that'll make you seem intellective to co workers and fellow commuters on the subway. But there's no chance that a little book like this could save your life. Weather is something that happens to impoverished people in Indochina and Seattle. It can't possibly get to you—you sitting there in your finely furnished living room inside your expensive weatherproof home, sipping perfectly aged whiskey with all your smug assumptions about this book. No, not you. You're safe.

Unless, maybe, you're not so much.

What are you going to do when a fierce bolt of lightning shatters its way through your living room's bay window and explosively bores a flaming gorge into your fine wall-to-wall carpeting? Most likely you won't know what to do, rendered deaf and bewildered as you most certainly will be by its cacophonous clamor. No, you'll wander confused and horrified through the dense black smoke that chokes your lungs and muddies your sight, vibrantly hot flames teasing at your limbs with cruel tongues, so painful indeed that you will almost welcome the deluge of rainwater that rushes into your home through its newly torn breach. That is, until the water level rises above your waist and your seared and weakened legs can hold your balance no longer and you go under. Below the waterline of the flood, in the quiet—the deep, deep quiet—you may find some peace. But no, actually you won't, because the winds are whipping up. Fierce northerly winds that chill the room and stir the water into great white crests that crash mercilessly about your beleaguered frame, knocking you this way and that. Carrying you with unfathomable force to the far end of the room and then, as the waves recede, back and under again.

And then there are hailstones. Big, giant, gigantic hailstones. As big as nonregulation-sized softballs. Yeah. And they come crashing through your roof and landing on your head. And they hurt. A lot! And then there's a tornado. Yeah, a tornado. It's not a very big tornado, because that wouldn't be able to fit inside your house, right? So it's a smaller tornado that fits inside your house. And it's all twisting around and stuff, and it picks you up and spins you around. And you get all dizzy and you want to throw up. But before you can throw up, something else happens. A glacier. Yes, a glacier. A huge frozen wall-of-ice glacier splinters your precious peaceful "weatherproof" home into toothpicks and encases you—screaming and horrified—beneath tons and tons of ice. And you stay frozen in place like that, a monument to disaster, for thousands of years until hikers find your petrified form and dig you out and eat you.

Safe? Ha! You think it can't happen? Well, it *can* happen. Probably not the part about the glacier. But everything up to that. Bet on it.

Weather is out there, and it's everywhere—outside your front door, hovering over your beach house, skulking above your local baseball stadium. There's no getting away from it. But, you may ask, *what can you do about it?*

Can this looming menace be destroyed? Unfortunately, the answer to that question is no. At least not in the short term, but we'll get into the theoretical destruction of all of Weather's effects on the planet as we know it in more depth later. For now let's stay focused on what we can do to hold back the torrents of Weather's torrential destructive torrents *today.*

So, are we to remain forever locked within Weather's cruel grip? Or is there some way that we can somehow presage its devious campaign, outsmart it, and beat it at its own game? The answer here is yes. Luckily for you, you recently acquired this nifty little book, which contains within it all the secrets of the fine art of weather prediction. It is an art that we will learn together over the course of its lifesaving chapters. Or perhaps you could place this book back upon the shelf, turn on the local news, and let some guy tell you what *he thinks* Weather has in store for us. While you're at it, why not give your dog the keys to your Lexus and ask him to go have the tires rotated? Get back to me and let me know how that all worked out. Meanwhile we'll be working things out on our own.

The first thing that we'll need to do is get inside Weather's mind. Try to understand how it works and why it makes the decisions it makes. There's a reason why the Southeast is bombarded with hurricanes in the early fall and the High Plains become pockmarked by hailstones in the summer. Why storms move one way in the Northern Hemisphere and another in the Southern Hemisphere.

Weather's Circuitous Dance of Terror

Let's take a step inside what some people call **The Hydrologic Cycle** but what is probably better described as **Weather's Circuitous Dance of Terror.** This is an ever-repeating circuit of energy and moisture transfer, and it is the central cog to Weather's diabolical mechanisms. There is no beginning, and there is no end; just when

Hydrologic Cycle

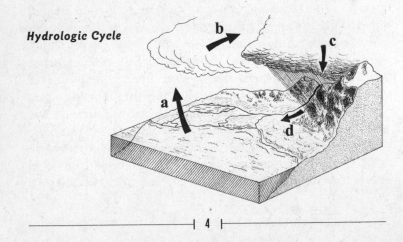

you think it's all over, it goes right back to its starting point. With that as such, there's no "best" place to jump into the cycle, so let's just choose a point at random . . .

The water molecules have now had enough, and they begin pushing their cloud droplets together with a bunch of other cloud droplets to make one massive raindrop. Now, finally, they've collected into something heavy enough to burst free from the cloud. They push their new massive raindrop through the cloud, picking up smaller droplets along the way, seeking free fall. Their raindrop gets bigger and bigger and . . .

You know what? That's not such a good place to start after all. On second thought, let's be a little bit smarter about this . . .

We'll begin our journey with a huge flaming ball of thermonuclear explosions we call the Sun. As its surface burns away at 27,000,000°F, it hurls radiation at our tiny planet some ninety-three million miles away. Of the tiny fraction of this ration that reaches our atmosphere, most of it will be swallowed up by the atmosphere. Only about eighteen watts per square foot—the approximate energy of eighteen commemorative Elvis candles burning down to sideburn length in perpetuity— actually finish the trip and collide with the 71 percent of our planet's surface that makes up the oceans, lakes, rivers, streams, puddles, and kiddie pools: water.

One little water molecule has decided that the ocean is a boring place to be. It's tired of being constantly pushed around by the oceans' currents. It wants to escape, so it pushes itself to the water's surface to catch the falling solar radiation, which excites its two hydrogen and one oxygen atoms. It revels in the feeling of its electrons moving faster and farther out in their orbits. The water molecule can feel itself expanding, and it breaks itself free from its fellow water molecules and leaps into the air above, climbing up upon the backs of the bajillions of denser airborne particles scrambling to get up into the sky. It climbs up and up and up. Alas, the little water molecule is still not quite light enough to make it all the way up. It seems destined to lose all its energy and fall back down to the bottom of the pile and into the ocean. But that's not to be for this water molecule; it's about to get a helping hand from the wind.

You see, because the Sun is shining down here, it means it's not shining down someplace else. The Sun can be shining only on one-half of the globe at any one time, and because the Earth is a sphere, sunlight is always going to be shining brighter on one spot on the sphere than it is on all the other spots. The sunlight here is making all the airborne particles' atoms much more jumpy and erratic here than the air molecules someplace else. So all these air particles are spread out and bumping up against one another while someplace else all the air particles are getting tired and denser and finding themselves with much more space than they need. Naturally all the air particles here decide that it would be a fantastic idea to rush as fast as they can over to the someplace else where there's much more space for them to be jumpy and erratic. And so they do, and now we have several tons of air particles all moving at once, acting as one force of wind.

And the little water molecule, oblivious to any of the chaos, gets carried along for the ride. It gets pushed along and up and over, and now it's on the top of a small pile of air particles. And then it's on the top of a large pile of air particles. And before it even knows what's going on, the water molecule is way up in the air. It's getting caught up in stronger and stronger air currents as it rises through the troposphere. And the temperature is getting colder and colder.

It ends up really high up in the sky, and without a whole lot of other water molecules to bump into, things are seeming a lot less exciting. In fact it's starting to think that it could be into the idea of condensing itself onto some piece of anything and getting back home to the ocean. But it can't get back down because anytime it tries, the wind keeps pushing it back up. And it's not alone. As it turns out, there are other expatriate water molecules floating around up here in the sky. So as expatriates often do, they start hanging around with one another. More water molecules join their group, and then even more water molecules join their group, and before you know it, our water molecule is part of a cloud.

From the ground clouds may look like puffy pieces of fluff, but up close they're densely packed ghettos of disaffected water particles, full of potential energy they're unable to expend. So they float around up in the sky, their water molecules forming small collectives called cloud droplets, waiting for their opportunity to escape back to the surface.

Meanwhile the winds are still moving, constantly shifting from one place to the next. No sooner do their air molecules arrive at one location than they decide that it's too warm and crowded here and they have to get someplace cooler and more spacious. And it's not from one part of the sky to another. Their travels span entire continents and oceans. Above the equator they skirt the tropics westward across the Atlantic and the Pacific and then swirl upward, blustering back east. And then up and west again. And by the time they reach the North Pole, they've cooled and calmed down sufficiently that they've lost the will to fight, so they get pushed back down southward by warmer, more excited air molecules. It's a tireless nomadic life for an air molecule. But it's not much better for a water molecule that finds itself caught up in the chaotic life of the sky. Water vapor is constantly getting pushed and pulled along with the winds. So the cloud with our water molecule is getting to see the world.

After about ten days and thousands of miles of travel logged, our water molecule and his angry refugee friends in the cloud want out. They've been conspiring for some time now, colliding into one another to form larger, heavier droplets. But they're still not quite heavy enough to break the tyranny of the wind that keeps pushing them back up every time they try to fall. Then they see an opportunity. The warm wind they've been traveling with has just encountered its archnemesis coming the other way, a cold wind that has been displaced by some other warm wind someplace else. The two winds meet head-on, each one unwilling to stand down from its opposed path of travel. As they clash, the cold air hunkers down low, trying to bowl through the warm air, which in turn seeks the high ground, choosing to fight from a place of strategic advantage.

In the course of battle the water molecules are shoved upward, higher into the much cooler sky, where they gather together to form larger and larger droplets. Their small puffy cloud becomes a large heavy cloud as other water molecules also forced into the upper sky join them. They become so thick and heavy that they cover the sky and blot out the Sun, making them appear black and dangerous to anyone unlucky enough to be viewing them from beneath.

The water molecules have now had enough, and they begin pushing their cloud droplets together with a bunch of other cloud droplets to make one massive raindrop. Now, finally, they've collected into something heavy enough to burst free from the cloud. They push their new massive raindrop through the cloud, picking up smaller droplets along the way, seeking free fall. Their raindrop gets bigger and bigger and bigger and, as it grows, gets faster and faster and faster. And they're not alone. There are suddenly huge raindrops everywhere, pushing down into the winds' fray, getting pushed right, left, up, and down as the two winds indiscriminately engage each other. Some of the raindrops don't make it. One powerful gust of wind can push an entire battalion of raindrops back into the sky. Other raindrops, if not quite large enough, will be reevaporated by warm air, decimated and defeated, broken back into tiny water molecules that have no chance at it on their own.

But the ones that do make it through hurl themselves forcefully, angrily toward the Earth. There's no time to worry about getting back to the ocean now. The water molecules just want to make it back down to the surface. Still pushed along by the wind, the raindrops move diagonally toward their target at breakneck speeds. They slam explosively into anything they can find—buildings, trees, billboards, people, skateboards, sidewalks—finally breaking their bond as raindrops and tumbling out into a fine layer of wetness. But more raindrops follow, adding volume to the water that has already accumulated, piling up into puddles and spilling over. The newly liberated water molecules run streaming down walls and flowing through the gutters of streets in tiny rivers seeking the low ground. And they just keep falling.

When the battle has ceased and what's left of the cold and warm winds have gone their own ways, the skies clear. The clouds vanish. Those droplets that weren't lucky enough to make the trip down have dropped to a lower warmer altitude and dissipated back into vapor, pushed and pulled somewhere else with the winds. With the barrier of clouds gone, the sun is now free to shine its warm radiation upon the ground. Many of the water molecules that haven't yet made their way to safety—a water drain, loose soil, a large body of water—find themselves evaporated all over again, pulled back up into the unforgiving sky, getting pushed along again by another wind.

But for the lucky water molecule, it is the beginning of a brand-new journey. Our water molecule finds itself racing downhill. It hooks up with a large puddle that grows to the point where it can spill over into a small creek. The molecule follows the creek through its twists and curves, being joined along the way by more molecules from the storm, until the creek becomes a stream. It follows the stream until it empties into a gigantic river, full of molecules that have survived other storms in other places all over the region, all of them rushing as fast as they can back to their home in the ocean. And when our molecule arrives home, it looks around and decides that the ocean is a boring place to be.

And that in essence is **Weather's Circuitous Dance of Terror** (aka **the Hydrologic Cycle**), a devious plan that

Weather has devised for using the Sun and the Earth to pit air vs. water in a continuous loop with the result being violence, destruction, and inconvenience for us humans.

In the coming chapters we'll take a closer look at some of the more important aspects of this process in an attempt to better understand the way that Weather manipulates these tools against us.

CHAPTER 2
DISSECTING THE ATMOSPHERE

What is air? We can't see it, but we can feel it as it slips between our limbs, tousles our hair, and knocks our half-empty cans of soda from outdoor café tables. It's incorporeal enough for our hands to pass right through yet strong enough to support the weight of birds and airplanes beneath their wings. It carries the sounds of music to our ears, the smell of urinals to our noses, and paper airplanes to our co-workers' desks. It fills up our lungs and pours from our mouths, which kind of sounds gross but actually is beautiful. And it's everywhere, completely enshrouding the Earth with its airiness. It is what makes up our **atmosphere.**

But what is the atmosphere exactly? The *Oxford English Dictionary* defines *atmosphere* as "the pervading tone or mood of a place, situation or creative work." That's true, sort of. But only in a very vague sense. The atmosphere is more like a layer of gases that envelops the planet, held tight to its surface by the force of gravity, and keeps global temperatures at a habitable level, while also protecting us from terrible, horrible, dangerous radiation from outer space. Without the atmosphere, everyone you have ever known, met, stood in line behind, or argued with over the telephone would be dead. Asphyxiated. Frozen. Desiccated.

The atmosphere is made up of approximately one hundred tredecillion (or one-hundred sextillion sextillion or 10^{44} or a one with forty-four zeros after it) air molecules. Every time you take in a breath, the atmosphere is shoving somewhere in the neighborhood of ten sextillion (or ten thousand billion billion or 10^{22} or 10,000,000,000,000,000,000,000) invisible air molecules into your lungs. In case you're bad at math, that's a lot. It's about the same number of air molecules as there are stars in the universe. That's how many air molecules you pull into your lungs *every time you take a breath.* And then, when you exhale, you're pushing ten sextillion air molecules back out into the atmosphere.

If there's so much of it all around us, why can't we see it? Luckily its molecules reflect visible light only very mildly, which is good for us, or else we'd have to watch television inside a vacuum chamber.

Let's get out our microscopes, grab a handful of air, and see what it looks like close up. That is, if we could in fact see it, which we can't. But we'll make it work somehow.

Components of the Atmosphere

Do you know how many different gases make up our atmosphere? Do you have any idea? I personally do not.

But I have a feeling that it's a whole, whole lot. Anyway, we're going to focus mainly on the important ones that people care about. The other ones can suck it.

Nitrogen

The Earth's atmosphere is made up primarily of nitrogen (N_2). Approximately 78.084 percent, give or take a thousandth of a percent. Not only is nitrogen by far the most prevalent atmospheric gas, but it is also by far the most boring. Really, considering that it constitutes more than three-fourths the air we breathe and exist in, you'd think it would do something at least mildly cool. But no. It really doesn't do much of anything aside from some incredibly tedious business involving the creation of amino and nucleic acids, which are supposedly some sort of essential building blocks for all living things on Earth or something. Oh, and it does some kind of thing like keep all the air in the world from blowing up. Apparently it's inert, and its diluting effect upon the atmosphere keeps the other explosion-prone gases from igniting into a worldwide apocalyptic ball of ignited apocalyptness. So there you go; it's a stabilizing force. Blah blah blah. Moving on . . .

Oxygen

The second most prevalent element in the atmosphere is oxygen (O_2), coming in at around 20.946 percent. Oxygen isn't exactly a combustible or explosionable gas itself. But it's very useful for helping other gases that are combustible and explosionable to combust and explode. Not only that, but it's also vitally important for the survival of practically every form of animal life on the planet because it helps them to not be dead. And you know what? It does both things for the exact same reason! *Kablow!* Did you just have your mind exploded and combusted? Were you at least intrigued? The hell with you.

You see, while oxygen does not itself ignite and burn as a fuel—in the way that gasoline does—it does *react* with fuels and *allow* them to burn. Without oxygen (or a couple other reactive agents) nothing, including gasoline, would burn. And no one would breathe.

In the same way that a commemorative Elvis candle uses the oxygen in the surrounding air as a catalyst for the burning of its wax, by which it unleashes the stores of potential energy held within, our cells use the oxygen that we inhale to burn and unleash the stores of energy contained in the food we consume. When a commemorative Elvis candle is lit, the oxygen surrounding it reacts with the carbon in the wax, initiating a somewhat complex sequence of chemical reactions, and part of the output is energy—in this case, heat and light. Pretty much the same thing happens inside our cells every time we take a breath. The oxygen in the air is carried by our blood from our lungs to cells all throughout the body. Inside the cell the oxygen is used as a reactive agent to help burn up the carbon in the cell's reserves of fat, protein, and/or sugar. Another somewhat complex sequence of chemical reactions occurs, and once again, part of the output is energy. This energy is what we use to think, move, eat, play chess, hunt squirrels, and breathe. And then the process starts all over again.

Oxygen is also necessary for the creation of water, carbon dioxide, and the lemongrass-infused oxygen cocktails we purchase at midtown upscale oxygen bars. All of which are very important to life on Earth.

Carbon Dioxide

Carbon dioxide (CO_2) accounts for a mere 0.0383 percent of the atmosphere. That means that for every million particles in the air, only 383 are carbon dioxide. That's not a lot. However, that little amount of CO_2 serves a very valuable function: you (assuming you are a person of some intrinsic value and not a K Street lobbyist). It's responsible for pretty much everything. Every human, animal, plant, and lobbyist are essentially made out of CO_2. And it all comes right out of the air. Sound preposterous? Well, feast your brain on this:

Plants absorb carbon dioxide in much the same way that our lungs absorb oxygen. They then—through photosynthesis—use energy from the sun, aided by water absorbed from the ground, to convert that carbon dioxide into carbon-based food, keeping the carbon bits and spitting out the O_2 (which animals in turn breathe). They use the carbon to build their cells and grow up strong and tall. Then a cow comes along and eats them. The cow uses the carbon from the plant to help *it* grow up strong and delicious. Eventually, when the cow has eaten enough plants and is strong and delicious enough, a person stuns, slaughters, butchers, grills, and eats the cow, using the carbon from the cow to help him grow up strong and smart enough to write a book about Weather. It's a perfect system. (If you happen to be a vegetarian, that's too bad for you. Why don't you go and fight the cow for that plant?)

So, plants use carbon dioxide as their building blocks. Animals and vegetarians use plants as their building blocks. And people use animals (and, in some cases, vegetarians) as their building blocks. Ergo all of us are made out of carbon dioxide that was just pulled out of the air.

To go further with this idea, anything of organic origin is made out of carbon dioxide. Your desk. Your shirt. That bagel. The gasoline you siphon out of your neighbor's car. Anything derived from petroleum, which includes plastic. All of it in a sense is made out of air. Check and mate. Lorem ipsum dolor sit. Case closed.

After that rousing victory, I'd like to take a quick U-turn and go back to oxygen for a second. With this new revelation, you can now see that what the oxygen is actually doing within our cells is reacting with the carbon we received from the plant and then facilitating a chemical reaction that lets off the energy that the plant absorbed from the sun. The oxygen (O_2) reunites with the carbon (C) and produces energy, which we use, plus carbon dioxide (CO_2), which we exhale back into the atmosphere.

And that's not all that carbon dioxide is good for. It's also useful for destroying the Earth's fragile ecosystem and hitting Weather where it hurts. (More on that later in the book.)

Argon

Argon (Ar) is an element that makes up slightly less than 1 percent of the atmosphere and is what is known as a noble gas, which means it does not react with other elements. Did we already establish nitrogen to be the most boring element in the atmosphere? Well, argon is even more the most boring element in the atmosphere. It doesn't do anything. Just like the British nobility.

Water Vapor

Around one-quarter of 1 percent of the atmosphere is made up of water vapor (H_2O), and as we've seen in the previous chapter, this is an extremely dangerous component of the air. (It's a major player in Weather's cruel game of death and inconvenience.) But we also know that it's kind of important for us to keep it around in case we ever get thirsty. Because humans are made up of approximately 112 percent water (or something like that), if we did get rid of it, we'd almost certainly shrivel up like strips of people jerky. On top of that, it's used by plants for photosynthesis. So we're sort of stuck with this one.

The Junk Components

The remaining 0.002 percent of the atmosphere is filled up by innumerable other crap gases and microscopic particles of airborne garbage. These include neon, helium, methane, hydrogen, krypton (presumably bits of Superman's home planet, blown here after its sad, sad explosion), ammonia, nitrous oxide, particles of dust, pollen, sea salt, dead skin cells, old pot smoke, cotton molecules from that old Ozzy Osbourne concert T-shirt you wore to death back in fourth grade, tiny little pieces of meteorite and long-dead stars, volcanic ash, and birds.

In concentration, any of these may have some kind of effect or worth, but spread out in such small amounts throughout the atmosphere as they are, they have no kind of effect or worth and can be ignored.

Ignore them.

Ozone

This is one of the few exceptions to the above components; it actually has an effect on our lives. Though ozone (O_3) is present within the atmosphere only in minuscule quantities and is virtually nonpresent on the surface, its value to us is immense. This heroic little molecule protects us from the sun's radiation. We'll get into that more so in the next section on how we can divvy up the levels of the atmosphere (which, as of the writing of this paragraph, will probably be called something like "Divide and Conquer, O Ye Brave Warriors!").

An interesting thing about ozone is that it's really nothing more than a fancy kind of oxygen. A molecule of oxygen is normally found as two oxygen atoms fused together (in the symbol O_2, that $_2$ stands for "two"). But ozone is *three* oxygen atoms fused together. That's cool because it's different.

Now that we know *what* air is, it's time to examine *where* air is. Where is air? It's everywhere, isn't it? We're in it right now, and it's in us right now. It's almost impossible to imagine a situation in which you would find yourself not completely immersed in air. You can travel to a location thousands and thousands of miles away from your home, and when you get there, you'll find air. There is no place that you can go where you will not find a plentiful supply of air.

Actually, there is one place you can go. *Up.*

Levels of the Atmosphere
(Formerly Divide and Conquer,
O Ye Brave Warriors!)

Let's imagine that you are standing on the Earth's surface. Somewhere roughly at sea level. Let's say that it's New Jersey. No, not New Jersey—I have family there, and I'm not in the mood to get into a big conversation with them about things. So, let's say Honduras? Okay, so you're standing on the Earth's surface someplace in Kansas. But to be honest, I don't know anything about Kansas; I don't even know where to get a decent steak there. But you know what? It doesn't matter because we're talking about you and not me. So, let's just make it New Jersey. But if you see my family, just don't mention me or this book or anything. You're not going to be there that long anyway.

So, you're on the Earth's surface in New Jersey, and you might not realize it, but you're standing beneath a humongous pile of air. It's piled way up on top of you, going up thousands and thousands of feet into the air. And you can climb out from under this humongous pile of air, and it won't matter at all, because there's another humongous pile of air over there waiting to get on top of you. Go wherever you want; you can even leave New Jersey and go to Kansas, and there'll be another humongous pile of air waiting for you. There's no getting away from it, because you and me and all your friends and my family and whoever it was who stole the iPod out of my bag at Starbucks all spend our lives at what is essentially the very bottom of an ocean of air.

It's pressing down on you from above, and it's pressing sideways on you from the side. It's pressing up on you from underneath. It's even pushing out of you from the inside. Which, by the way, is why you're not dead, squashed flat into a pink and orange stain on the New Jersey ground. Because humans—and cats and dogs and marmosets, if marmosets are in fact animals and not a pastry—are built to live at the bottom of this invisible ocean. Even the strongest, biggest, most bitchin'est birds in the world barely leave the bottom of this ocean of air, relatively speaking. Depending on what you con-

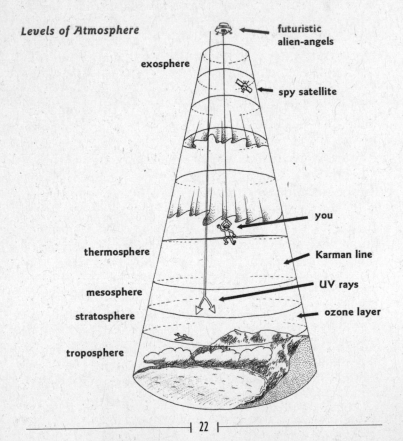

Levels of Atmosphere

futuristic alien-angels

exosphere

spy satellite

you

thermosphere

Karman line

UV rays

mesosphere

stratosphere

ozone layer

troposphere

sider the "top" of the atmosphere to be, this ocean extends anywhere between sixty-two miles up to somewhere out in space. The strongest birds barely even make it six miles up. That's about fifty-six miles below even the most conservative mark.

The Troposphere

This ocean is divided up into five separate levels, each with different properties. The bottommost level in which we all live is called the **troposphere.** The word comes from the Greek word *tropos,* meaning "turning" or "mixing," and the word *sphaira,* meaning "round, bally thing." That second part is kind of stupid, but the first part actually makes sense, because this is the level of the atmosphere in which all the shit goes down. Different kinds of air (dry, moist, hot, cold, polluted, perfumed) are constantly *mixing* in with one another and *turning* into different kinds of Weather. (Wow. That actually kinda made sense.)

But let's get back to what we came here to New Jersey for. You're standing beneath a ton of air at the bottom of this invisible ocean and blah blah blah. Now, look up. What do you see? You probably see a lot of clouds of different shapes and sizes (the categorical names of which are not important just yet). Maybe a news helicopter or an errant balloon. Maybe, if you're lucky, one of those football zeppelins with a funny phrase spelled out in lights. Look higher. What's that at the very top?

If you said *the sky,* then you're wrong. The sky is not the top. In fact the sky, vibrant blue as we see it, is very far from the top actually. Our atmosphere goes up miles above. The blue color you're seeing is really closer to being the bottom than the top. When light arrives at the Earth from the Sun, it's white light, meaning it still contains all the colors of the rainbow. But once that white light collides with an air molecule in our atmosphere, the blue part of the white light gets absorbed by the air molecules and then radiated out in all directions—one of those directions being into your eyeballs—while most of the rest of the white light just passes on through and continues on its way. Why does the rest of the light pass through while the blue is absorbed and reflected? That's an excellent question. And there's a really interesting answer. And I would strongly encourage somebody who's writing a book about the properties of light to include that information. Anyway, most of the blue that you're seeing is being reflected from collisions happening not all that many miles above our heads because as you'll see in a bit . . . Well, you'll see in a bit.

Hey, what's this thing coming down from the sky? Why, it's a cord. That's crazy. I had no idea something like this would happen right now here in this chapter. Looks like the cord's descending just above you. Who do you think could be lowering that? Maybe it's aliens, or maybe some angels. Maybe they're dead aliens who are now angels. From the future. Whatever. It doesn't matter, because this is a metaphorical cord. And at the very end of this cord is a metaphorical clamp. Like a robotic clamp. Like in those machines in which you try to pick up stuffed animals at the arcade. Except in this metaphor, you're the stuffed animal and the futuristic alien-angels are attempting to pick up *you*. And they're superevolved or have angel powers or something, so they get you right away on the first try. The clamp is coming right down onto your head. And now that it has you, it's going to start pulling you up, up, up into the sky.

As you rise, suspended by your head, the first thing you'll probably notice—aside from the intense pain in your neck and spine—is that the air seems to be getting colder. You might think that since you're getting closer to the Sun, the air around you would get warmer, but that would be laughably naive. I want to laugh at you. The relative distance between the Sun and the Earth's surface and the Sun and a few miles up from the Earth's surface is negligible to the point of being nil. No, the main reason that air gets colder as you get higher up is that the Sun's radiation heats the Earth's surface, and the Earth's surface in turn radiates heat into the air around it.

This altitude-related drop in temperature is called the **lapse rate,** and it generally accounts for a 3.6°F drop in temperature for every thousand feet you rise. (However, for reasons science does not yet understand, this equation holds only in the United States; elsewhere the lapse rate is 6.5°C for every thousand meters.) The lapse rate is not a hard-and-fast rule. It changes somewhat depending upon latitude, season, and whether or not you're ascending from an active volcano. In some situations, such as when a warm air mass rolls over a cooler one, it may even get hotter as you rise. This is called **temperature inversion,** and it's not uncommon; but it's also not the norm.

If we assume that the air temperature back at the surface in New Jersey was around 80°F, after you've climbed one thousand feet, the temperature would drop to around 76°F, and at around the mile mark (a little more than five thousand feet up), it would have dropped to 62°F. And by the time you travel the six miles (or 31,680 feet) up to where the troposphere ends, the temperature will have dropped to –35°F.

Another thing you may notice as you rise is that the air is getting thinner, and it's harder for you to breathe. Gravity constantly pulls as many air molecules to the Earth's surface as it can. So you'll find a lot more air molecules crammed together near the ground and therefore a lot more heat transfer as well. That's another reason the air gets cooler as you rise. The higher up you go, the fewer molecules you'll find slamming into one another and transferring their energy from molecule to molecule to molecule. Not only that, but the molecules are flying all around, expending energy, and by not hitting other molecules that would give it energy, it's actually losing energy and becoming cooler.

Not only does the air get thinner as you rise, but it gets exponentially thinner. Once you've been pulled a mere three miles up into the sky, a full 50 percent of the Earth's atmosphere is beneath you. And as the futuristic alien-angels pull you up past all the different kinds of clouds (names still not important yet) to the top of the troposphere, and the outline of New Jersey is lying like a tired, embittered postal worker on the ground, roughly 85 percent of the Earth's atmosphere is whooshing (or swishing, depending on the season) beneath your dangling feet. You've traveled like only six miles; that's less than the distance most people travel to work.

At this point you've reached the **tropopause,** the barrier between the troposphere and what lies above. *Pause,* in this sense, means "to stop for good." It is the end of the troposphere, the point at which the lapse rate breaks down and temperature ceases to decrease with height. It's where the warmer temperatures—both below and above—even out, which keeps the air molecules, which always want to go from warmer to cooler areas, from rising any farther.

This is where Weather as we know it stops. Air molecules are rarely able to pass through this wall in the sky. The highest clouds will usually bump up against the troposphere and stop rising, flattening out into anvil-shaped clouds that vaguely resemble anvils in an anvily kind of way. You, however, will be going right on through.

The Stratosphere

The **stratosphere** is what lies directly above the tropopause. It derives its name from the Latin word *stratus* for "spreading out." So spread out and make yourself comfortable. This leg of your journey upward will be a little longer than the last.

You may notice that the air is considerably drier up here. That's because there's so little mixing of air between the troposphere and stratosphere. Water exists primarily on the Earth's surface, feeding the air nearest it; there aren't any lakes fifteen miles up in the sky, at least none that we've found so far. So because the moist tropospheric air can't push its way through the tropopause, the stratosphere stays pretty dry. Sure, some particularly strong storm clouds or incredibly forceful rivers of air (**jet streams,** but more on them in subsequent chapters) may pound their way through, but those are the exceptions, not the not-exceptions.

You may also notice that it's magnificently boring. There's nothing to look at. But that's good. Boringness means Weatherlessness. No clouds. Calm skies. Terrible kite-flying conditions. Commercial airlines regularly fly up here to avoid the unpredictable and sometimes dangerous Weather below. A plane full of happy, contented passengers could in theory fly right over a terrible, cacophonous storm and not even notice, if anybody was ever happy and contented on an airplane. That is life without Weather; it's the life for which we strive.

Now that you're ten miles up, 90 percent of all air molecules are beneath you. The temperature's steady right now, but it should begin to rise right . . . now. Temperature inversion. Here in the stratosphere, that's the norm. It's because of the **ozone layer,** which you're rising through right now. You might recognize the smell, like electric sparks, impending rain, or a bumper car rink. You entered it about a mile ago, and you should be in it for another eleven, so that's between the nine- and twenty-one-mile marks, although it varies depending upon latitude and season.

Here is where ultraviolet light from the Sun comes careening into oxygen molecules (O_2), knocking them into two free-floating oxygen atoms $(O + O)$. These two atoms get very lonely and clingy and go off to cling to whatever oxygen atoms they can find, which is usually an oxygen molecule, and an ozone molecule is born $(O + O_2 \rightarrow O_3)$. One particle of ultraviolet (UV) radiation will turn three oxygen molecules into two ozone molecules. This is happening constantly, and it's why the ozone layer exists in the first place. This twenty-or-so-mile strip contains the vast majority of all the ozone molecules on the planet, around 90 percent. Even there they account for less than ten in a million air molecules. But without them, a particle of ultraviolet radiation would have a much easier time making it all the way down to the surface and hitting a human cell.

There are three kinds of UV light, each existing at a different point on the light spectrum and as invisible to our eyes as street solicitors for Greenpeace. UV-C, with the shortest wavelength, is extremely hazardous to human cells; it gets into our cells' DNA and causes genetic mutations and cancer. UV-B, with a slightly longer wavelength, isn't so great either; this is the stuff that causes suntans and sunburns. UV-A, with the longest wavelength, isn't really dangerous at all; in fact it's rather useful for making our Grateful Dead posters look all trippy.

When ultraviolet light meets the ozone layer, the ozone molecules absorb those with the shorter wavelengths. UV-C particles are the first to go; they're completely neutralized before they ever have a chance to reach the surface. The UV-B is next, wiped out almost completely; some of it does sneak through, though most of it settles in the faces of movie producers and jewelry salesmen. UV-A travels through without incident.

What happens is an ultraviolet light particle hits an ozone molecule and the energy is expended in splitting it back into one oxygen molecule and one free-floating oxygen atom ($O_3 \rightarrow O_2 + O$). Then that free-floating oxygen atom gets lonely and clingy again and goes off to cling to another oxygen molecule and turn into ozone again, while the UV particle is destroyed, its energy sucked up by the ozone as heat energy, which is why the temperature rises with altitude up here. The higher up you go, the more UV particles you'll find careening into ozone and oxygen molecules. But closer to the tropopause, the UV light particles that would have reacted have already been killed.

As you continue to rise up and out of the ozone layer, continuing upward, the temperature will rise with you, even though you've left the bulk of the molecules and their heat-generating reactions behind. That's because the air is continuing to get thinner, so a few ozone/ultraviolet light encounters go a lot further. Plus a lot of the Sun's energy gets absorbed by air molecules here in the upper stratosphere before it ever makes it down to the lower stratosphere or troposphere. Remember how we were discussing where the blue in the sky comes from? It's here.

You've now come to the top of the stratosphere—the **stratopause**—about thirty-one miles above the Earth's surface; the temperature steadies out at about 26°F, and you're above 99.9 percent of the atmosphere's molecules even though you're less than 8 percent of the way up to its outermost layer. It may seem odd that such a small percentage of the atmosphere contains such a large percentage of its air. But maybe if the other 92 percent of the atmosphere would just work a little harder, it might accrue more air and not just piss it all away into outer space. Don't be fooled: Equal distribution of air is atmospheric socialism. And it simply doesn't work.

The Mesosphere

If one were to think of the short six miles of troposphere—with its 85 tredecillion (8.5 × 10^{43}) air molecules and immense power—as the wealthy elite and the twenty-five miles of stratosphere—with its 1.49 tredecillion (1.49 × 10^{43}) air molecules and invaluable ozone layer—as the upper class, then the cold, desolate, wispy **mesosphere** must be the middle class. In fact *mesos*, from which the layer gets its name, is Greek for "middle."

Most of the remaining 0.01 tredecillion (10^{41}) are contained within this barren twenty-mile band. It is largely unexplored and forgotten about, as its air is too thin for aircraft to fly through and its gravitational pull too strong for satellites to orbit within. It represents the end of the atmosphere as we know it and dangles precariously over the precipice of outer space. It's suffocating, stultifying, and hypothermic. Just like middle-class America.

As you're pulled upward, you'll find the temperature holding steady at about 26°F for a few miles, before it starts to drop again. Starting right . . . now. Wow, it's dropping quickly, huh? Yikes. (Hope you remembered to bring a jacket.) Oh, and watch out for that radiation. The UV particles won't get absorbed until they enter the stratosphere, and if any of them hit you (which they will), it *will* cause severe burning to exposed skin, genetic mutation, cancer, and sterilization. (Hope that jacket is coated in Mylar and covers your entire body.) Um, you should probably also watch out for your blood boiling. The air is really thin up here, and what with the boiling point for liquids dropping with decreased air pressure, your own body heat will most likely make the blood boil right inside your veins. (Hope that Mylar-coated full-body jacket is adequately pressurized.) And while we're on the subject, you're probably going to asphyxiate and die since you've got less than 0.1 percent of the available air you had back at the surface. (Hope that Mylar-coated full-body pressurized jacket has an oxygen tank in the pocket.)

But if you're not too severely burned, asphyxiated, sick, or dead by this point, you might want to take a look around and notice all the shooting stars. Yep, this is where most of the refuse from space gets burned up from the friction of hitting the relatively few molecules to be found. Of the millions that enter every day, the vast majority are tiny, no bigger than a grain of sand. They burn without much fanfare. But the larger ones—the pea-sized and orange-sized ones—they're something. Too bad only one in twenty thousand can be seen from the ground. But up here the shower of shooting stars is a pretty spectacular thing to see, huh?

Hey, you're not dead, are you? 'Cause you're fifty miles up at the top of the mesosphere, and you're getting ready to pass through the **mesopause.** (Hope that Mylar-coated full-body pressurized oxygen-fed jacket is a space suit.)

The Thermosphere

You are now entering the atmospheric ghetto known as the **thermosphere.** If you were uncomfortable with the chilly temperatures of the mesosphere, don't worry, because it's about to get considerably warmer. You may have already guessed that the Greek root *thermos* means "heat." At the mesopause the temperature evened out at around -146°F, it'll hold steady for a few miles, and then it will start to rise. Starting . . . now. Do you feel that spike in temperature? No? Oh, well, I guess that's because there's so few air particles up here that you're not going to run into them very often at all. But if you did, you'd find them to be a comfy 2,730°F. But don't worry, they cool down by about 1,000°F at night.

It's funny. Down on the surface no two air molecules are ever more than three-millionths of an inch away from each other. But up here any two molecules can be separated by as much as a mile, so there's very little heat transference. I like to think of it as a game of billiards, except with balls the size of viruses and a table the size of the Grand Canyon. Not my kind of game.

Hey, look at this invisible, theoretical border of some obscure importance coming up at sixty-two miles up. That's the **Karman line.** It's unofficially (but officially [but unofficially]) the official dividing line between Earth and outer space. It represents the point at which the atmosphere becomes too thin for nonrocket or superbeing flight purposes. Therefore anything above this imaginary line may be considered extraterrestrial. And you're leaving the safety of Mother Earth . . . now. Hey, look at you; you're an astronaut.

That border is also sometimes called the **turbopause**—the dividing line between the **homosphere** and the **heterosphere**—with a slightly more meteorological connotation. Back in the homosphere, where we just came from, the various gases of the atmosphere get continually mixed up and sloshed around together and end up with a somewhat consistent composition (except for water vapor and ozone), but here in the heterosphere the different gases settle—like the multiple layers of salad dressing that's been sitting on the shelf for a while—into separate strata, according to their chemical weight. So, in the lower heterosphere you'll find the heavier molecules, such as oxygen and nitrogen, and the lighter molecules, such as helium and hydrogen, higher up.

Are you starting to get the feeling that you're coming to the outer edge of the atmosphere? Well, you're not even close. Despite the fact that you're now unofficially officially in outer space, the thermosphere continues upward for at least another two hundred miles. But what's a couple hundred miles between astrophysicists? So relax and keep your eyes open for satellites. This is the altitude at which they begin to orbit.

. . .

Not for nothing, but the temperature should be about 4,500°F by now. No, I haven't seen a molecule either.

. . .

It's really boring up here.

. . .

. . .

I can make my face do this really weird thing. You wanna see?

. . .

. . .

. . .

Oh, look! You see that? You see that nothing up there? Yeah, that nothing right there at three hundred miles up from the surface. That place where there's not anything at all. That's it! That's the beginning of the **exosphere!**

The Exosphere

Congratulations! You've reached the outermost level of the Earth's atmosphere. The name is derived from the Greek word *exos* for "outside." It begins about three hundred miles up and has no technical end point. It just gets wispier and wispier and wispier until it is no more. I mean, yeah, it's pretty wispy now, but hey, scientists have to draw lines. That's what they get paid for. Anyway, it's at this point at which atoms and molecules that have reached **exit velocity** leave the Earth's gravitational pull forever and ever and ever and go off into the farthest reaches of outer space.

And that's it, we're done. Hope you enjoyed it. And you don't need that cord anymore. At this altitude you can just safely float. You won't fall, but you probably won't be able to break free of the planet's gravitational pull either.

Oh, the futuristic alien-angels? Yeah, I decided that was a stupid idea. Dumb. So I didn't bother to write them. No, it's just you and me. Here in orbit. But the thing is I gotta go prepare for the next chapter. It's gonna be *good;* it's about wind. So I'm gonna take off. But, uh, you can find your way down, right?

'Bye.

CHAPTER 3

EVERYTHING YOU NEED TO KNOW ABOUT WIND

Put simply, wind is nothing more than air attempting to get from an area of high pressure to an area of lower pressure. The greater the difference in pressure between the two areas, the faster the air will attempt to get there. And because the warm air of the equator helps create an area of high pressure there, while the cool air of the poles helps create areas of low pressure there, air, when viewed on a global level, wants to get from the equator to the poles. However, because the Earth is rotating, the air is forced to move in a curved path (right above the equator and left below it), which causes three distinct cells of air movement on each hemisphere. Two of these create winds that move continually toward the west, and one creates winds that generally move toward the east. And in the gaps between the cells you'll find areas of either extreme high pressure or extreme low pressure in which wind cannot easily break through. At the top of these gaps, air collects and gets funneled into very fast-moving circuits of wind that whip haphazardly across the Earth far above our heads.

And that's everything you need to know about wind!

CHAPTER 3.5
MORE
EVERYTHING
YOU NEED TO KNOW
ABOUT WIND

According to a certain editor for a certain publishing house that will remain nameless, the previous chapter did not adequately live up to its title. I don't know. I thought it was good. But, after many e-mails with many CAPITAL LETTERS and one particularly chilly telephone call, it has been decided that we would add this chapter as sort of an addendum. Just to clarify a few of the more confusing points and bulk up the word count.

Anyway . . .

Let's begin by taking a look at how Weather manages to move air from here to there and back again. Here is a column of air that reaches from the ground up to

Why Air Moves

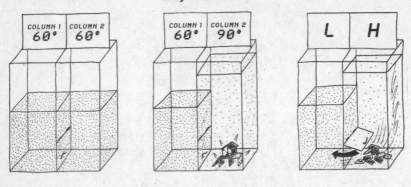

around a hundred feet in the air. You can't see it? Well, take my word for it; it's there. Oh, and there's another one just next to it, but separated from the first column by a pane of glass. Each one has the same number of molecules in it, and they're both pushing down with the same amount of force. This is the **air pressure** or **atmospheric pressure.** Right now they're the same temperature. About 60°F. Are you happy with that? Good.

Actually, that's a little too brisk for my taste (I'm more of a summer person), so you can have the first column with its autumnal temperature, and I'll take the second. I'm going to heat mine up to a nice cozy 90°F. (And I'll make myself a margarita while I'm at it.) Ah, this is the way to discuss Weather.

Look, something interesting is happening. My column of air is getting taller; it's dwarfing your puny autumnal column. As the molecules in my summertime column heat up, they're expanding and darting around faster and farther. They need more space, pushing and prodding all the other molecules around them. My column has a higher pressure than yours, so it grows to accommodate the increase in pressure. But the funny thing is, even though my column is so much bigger than yours, mine being **high pressure** while yours is **low pressure,** they're still exerting the same amount of air pressure onto the ground; we'll call that **surface pressure.** They still have the same number of molecules—and therefore the same weight—right? Right. It just stands to reason. So, just be happy with your column and find something fun to do. I'm going to build a house of cards in mine. (Why? No reason. It's just a hobby of mine that I do purely for fun and not at all to illustrate points.)

Wow, this is gonna be the best house of cards ever. It might even make it into the *Guinness Book of World Records.* Hey, what are you doing? Don't mess with the glass pane barrier. Listen, you chose the cooler column, so don't get mad at me now because I'm so happy here in my great big high-pressure Jimmy Buffett column and you have to wear a sweater in your little low-pressure Björk column. Stop it! You're gonna—

Damn it! Look what you did. By getting rid of the glass between the two columns, you're making all my air molecules go swooshing out of my column as they hurry over into yours. No, not my house of cards! Damn those treasonous air molecules! Now you've got more air molecules than I do, and your surface pressure, air pressure, and temperature are rising while all mine are sinking. My column is even shrinking in height; it doesn't have as many molecules to accommodate anymore. Air pressure, air density, and air temperature all are slaves to one another. When one changes, the others will as well. I'm gonna go get a jacket.

See, you fell right into Weather's trap. It pitted our two air columns against each other. *Air in high-pressure areas always seeks to find areas of lower pressure.* They always want to have more space to freak out and bounce around. And who paid the price? My poor house of cards did, that's who. By taking away the barrier between the two columns, you created wind.

Don't feel too bad. This kind of thing is happening constantly all over the world and on a much grander scale. There's always someplace somewhere that has a lower surface pressure than someplace else. So air is constantly trying to get someplace else. It's never happy with the place that it's at. It can't help it; it's a compulsion. So it's constantly moving. And that's all that wind is: air trying to get from someplace to someplace else. Sometimes faster and sometimes slower. The bigger the difference in air pressure between two points, the more forcefully it moves. That's called the **pressure gradient force.** So sometimes it's a gentle breeze caressing your cheek at an outdoor jazz festival, trying to get from one side of the park to another with a slightly lower pressure. And sometimes it's picking up your car and carrying it to the next town over, where there is considerably lower pressure.

There's a bunch of factors that go into the movement of wind, but the biggest, no doubt, is the disparity of heat throughout the globe. The sun can be shining on only one-half of the globe at any given time, so there's always going to be one side that's warmer than the other. Plus, because the globe is curvy and globy and not flat like a plane, the sunlight isn't going to hit all areas with the same intensity. Areas around the middle—the equator—get hit with much greater intensity than areas at the top and bottom, where the radiation gets diffused across much larger areas of land. As a result, the top and bottom are a lot cooler and therefore end up with considerably lower pressure.

We've already established that the atmosphere is like an ocean. Now try to consider the way that water moves around in an ocean. You never see a wave just hovering beside a trough. Whenever a spot of water is higher than another spot of water, the higher spot will slide into the lower spot. That's exactly the same thing that's happening all the time in the atmosphere. There are these waves of air that are forever splashing down onto the lower spots. In warm places like the equator, the air grows in pressure and height—like a swell in the ocean or my column of air—and then immediately seeks to move into the places with lower pressure and lower height.

And man-oh-man, does the high swell of equatorial air want to get to the lower polar air! It wants to get there so bad that it can just taste that sweet, sweet low-pressure awesomeness. So why doesn't it just go there and even everything out? There should just be one wind that blows constantly from the equator to the polar caps. You can predict that kind of wind. Well, it turns out it's not so simple. Weather is much more devious than that.

The Creepy Bed Effect

This is why you need to respect Weather as a formidable opponent. It's very clever. What it does is utilize the rotation of the Earth to send the wind spinning off to the right as soon as it gets going. This is called the **Coriolis force** or **Coriolis effect,** named for Gaspard-Gustave Coriolis, a French guy from the nineteenth century who liked to talk about how things rotate.

This is going to be difficult to explain, so let's make

ourselves comfortable. I just got one of those round, re-volving, satin-sheeted beds that the ladies find so irre-sistible. The guy who installed it wanted to make it rotate clockwise, but I said no. I prefer it to feel as if time is going backward when I sleep. Never mind the ceiling mirror; I like to comb my hair in bed. Don't look at me like that. Look, I'm not going to touch you. Just stay on your side of the bed, and I'll stay on mine. Just climb up so we can get this over with.

Wait, don't sit on the remote! Now you've got us spinning. I don't know why it's going so fast; there's something wrong with the motor. Turn it off! Hit the button! Quick, before we get sick. What do you mean you can't figure it out? Just toss it to me. That's right, just a simple straight toss.

That wasn't even close! It went way off to your right. What do you mean, you threw it in a straight line? It looked curved to me. Can't you just throw a remote like a regular person? Screw it! Just jump off.

Damn it! Now I've lost my concentration. If there were only some easy, not-at-all-hammered-in way to illus-trate the Coriolis effect. But clearly there's not. That de-fective counterclockwise-rotating bed and your inability to throw the remote in a simple straight line have put me all off. Hey, wait a minute . . . That reminds me of something . . .

I still have the receipt! I'm going to return this bed and get my money back.

Yes, of course I realize that the Earth rotates coun-terclockwise. So what? Well, yes, of course I understand that since we were on the bed while it was turning, the remote would appear to me to curve to the right, even though to anyone watching from outside the bed—like, perhaps, from the ceiling mirror—it would appear to have moved in a straight line while I turned out of its path. What are you getting at?

Stop trying to change the subject. Illustrating the Coriolis effect with some overly complex anecdote was just a bad idea on my part. I'm an idiot, okay. Let's just move on to how it affects the wind.

Anything that is free-moving—that is not fixed to the Earth's surface, via friction, nails, or superglue—is affected by the Coriolis effect. That goes for water, birds, baseballs, and, yes, air. Usually, it's way too subtle for us to notice or take into account, but it's happening all the time. In the Northern Hemisphere, all these things deflect to the right, and in the Southern Hemisphere, where they do everything wrong, things deflect to the left. When something—anything—leaves the ground, the Earth continues rotating beneath it. As long as it's fixed

Coriolis Effect

on the surface, it will rotate with the Earth, but once it leaves the surface—even by a few millimeters—all bets are off. The Earth isn't going to hold up its rotation and wait for it to come back down. So anything airborne is constantly curving.

This effect is most pronounced near the North and South poles, where the spin of the Earth is at its most shallow. This produces very tight curves. Much tighter than you'd find just twenty or so degrees lower in latitude. As you make your way closer and closer toward the equator, the effect continues its gradual loss of sway over the winds, producing ever-widening curves until it finally flatlines at the equator, the only line of latitude on the planet that is completely immune to the Coriolis effect. However, all you have to do is move either one foot north or one foot south, and you'll encounter Coriolis again, even if only to a minute degree.

A Series of Inner Tubes

Okay, so you've got some air *here*, and it wants to get across town to *there*, where the air pressure is a little lower. It attempts to blow its way over from *here* to *there*. But because of the Coriolis effect, it ends up slightly to the right of *there*. So what? Why should that matter to me at all? I don't care about that air. That air means nothing to me. Nothing.

Be that as it may, it should matter. To all of us. Sure, a little misplaced, confused air isn't the end of the world. But when you apply the Coriolis effect on a global level, its repercussions are massive. It is, in many ways, the reason that Weather is able to torture those of us who live in the middle-latitude areas of the world, such as most of North America. But we'll get into more on that later.

Right now we should discuss what happens when air attempts to move from the high-pressure equator to the low-pressure poles. Should be easy, right? It's a straight line from here to there. Except that it's not a straight line when the Coriolis effect is factored in. Now, all of a sudden, it's a never-ending series of curves and swirls and curlicues.

Here's the way air should move back and forth between the equator and the North Pole: The warm air of the equator should rise up to the top of the troposphere and then begin traveling north over the top of the cooler upper-latitude air beneath it. By the time it gets there, it will have cooled enough that it will descend to surface level. But then it will have to move southward along the surface and away from the North Pole to make room for the air that was moving along behind. It should keep moving southward like that as a continuous stream of equatorial air follows the same path that it did until it finds itself back at the equator again, at which point it starts the process all over again. There you go. One cell on each hemisphere, one connecting the equator to the North Pole and the other connecting it to the South Pole. A single continuous loop back and forth between the two points. Easy peasy.

But it's not easy peasy. That's not how it works at all. Because of the Coriolis effect, we have three cells per hemisphere, instead of one, the first and the third moving as our hypothetical cell did, with the air on top moving north and the air beneath moving south. But now, in between those two cells, we have a cell that moves in the opposite direction, with the air on top moving south and the air beneath moving north.

Why would that happen? How could the Coriolis effect create a Frankenstein system like that? You're just about to find out because you're going to make the trip.

Up, Up, Up

To follow the flow of air from the equator and up to the North Pole, we're going to need you first to go on a really strict low-carb diet so that you lose enough weight that you can actually get blown away by the wind.

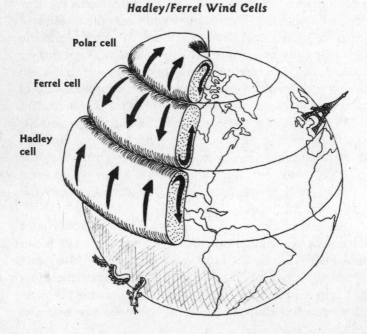

Hadley/Ferrel Wind Cells

Polar cell

Ferrel cell

Hadley cell

I'll wait.

Ready? Great. What took you so long?

Let's get started in Pontianak, Indonesia, on the island of Borneo, located precisely at 0° latitude, right on top of the equator. This is the spot where the sun beats down with its strongest intensity; it's hitting the equator directly, and not on an angle, so its radiation isn't getting diffused or scattered across a whole lot of land.

So it's hot. The air gets hot. When the air gets hot, its molecules get all jittery and spread out, looking for space, so they begin to rise (just as they did in my column, before you stole them). And you, light as you are, move with them. It's just like being on a hot air balloon, except without the balloon. It's like being on a hot air hot air. You and the air rise up until you reach the tropopause, and once there you can't rise anymore. (Remember from Chapter 2 how the lapse rate breaks down?) What happens now? The air can't go up any higher, but there's a bunch of other air beneath it, pushing its way upward. So it spreads out, like a dollop of whipped cream if you squashed it against your ceiling with your palm for some reason. Some of it goes north, and some of it goes south. You're going to go with the north-moving air. (Everything that you experience will be mirrored by some alternate reality version of you that went south.)

The Hadley Cell

You're moving into something called the **Hadley cell,** named for an English person who tried way back in the eighteenth century to figure out what's about to happen to you. A cell of air is like a big puffy inner tube of air that is constantly rolling forward toward the poles (in the Northern Hemisphere, northward at its top and southward at its bottom; opposite in the Southern Hemisphere), like a conveyor belt. You're getting pushed up and over.

But remember the Coriolis effect that I unsuccessfully attempted to explain on my rotating bed with the remote. So, while you're moving northward with the air, you're also going to be moving to the east. You're going to ride this cell up to around 30° latitude, by which time the air around you will have cooled enough so that it will begin to fall once again over the tropics.

But something else is happening while you're aloft. All those air molecules that went up and over left a gap of low pressure beneath them. Air a little more to the north that already went up and over isn't going to let an opportunity like that go to waste. It rushes southward to fill the space.

But again the Coriolis effect is in effect, and as the air moves south, it's also turning toward the right—in this case, toward the west. And then you and your air are going to come down and do the exact same thing, and before you know it, you're caught up in a band of strong southwest-moving air at the surface level.

These are the **trade winds,** a huge wind pattern that circles the world. Sailors used to hitch a ride on its gales to power their ships from Africa to the Americas, and that's where its name comes from. *Trade* is a Middle English word meaning "path."

Are you confused? If so, I'm sorry to drop this on you now, but it's unlikely there'll be a better time, and you'll need to know this sooner or later. Those, uh, winds with which you're moving westward: Those are called **easterlies.** Winds are always classified by where they're blowing *from,* not *toward.* So anytime a withered old New Englander whittling on his front porch warns you of an impending nor'easter, he's talking about a storm blowing *from* the northeast *toward* the southwest. Just take a moment to wrap your head around that.

Back to business. You're caught up in these easterly trade winds, which are generally quite as innocuous and predictable as a *New Yorker* cartoon. You can almost always count on them to blow in the same way they always blow: calmly from the east. This makes for some of Weather's less unappealing conditions. Eventually you'll make your way far enough south to hit the equator again and then start the whole process over. When you do, you may notice how much more calm the air is down here than it was in the trade winds. Because the air is so hot here, the air pressure is extraordinarily high. Therefore there is not a whole lot of air coming *into* this area. Why would air want to? It wants to get out, and it does. It goes up and over, just as you did. But that makes for not a lot of wind at the surface level, which in turn makes for poor sailing and kite-flying conditions (but excellent card house–building conditions). These are the **doldrums.** You'll understand that name all too well should you ever find yourself caught on a sailboat down here. Like the Hadley cell, they stretch all the way around the globe in the very low latitudes.

If not for the rotation of the Earth and the Coriolis effect, this one cell would stretch all the way up to the North Pole, creating two cells on either hemisphere. Air would rise at the equator, move north (with no turning right or anything) and sink at the pole, and then slide back south to the equator. That would make everything much easier. But no. No such luck.

Instead, at around the Tropic of Cancer at 30° latitude, you're going to come across a high-pressure band of inactivity similar to the doldrums. If you can slip in there on your next time around this rotisserie of air, you'll get to experience the **subtropical high.**

This area is no fun. No fun at all. Many a sailor unlucky enough to drift into this band found himself staring blankly at his ship's flaccid sails for days at time. Praying for anything. Just a simple little breeze. But nothing.

Over time this came to be known as the **horse latitudes.** According to legend, desperate seafarers of days past who drifted into these latitudes would seek to lighten their loads and drift more easily right back out, so they'd look for something heavy they could push overboard. For some reason, the horses, which were apparently employed as crewmen in those days, always seemed to lose the coin toss.

And the situation isn't any better on land. Because these latitudes have such a high surface pressure, there's not a whole lot of water making its way in. Wind has no interest in getting in, so storm clouds don't get in either, so it's rarely irrigated. Consequently most of the world's deserts lie somewhere around 15° to 30° latitude. (Most desserts, on the other hand, lie at around 50° latitude.) If you were to look at a map, you'd notice that the Sahara and the Mojave Desert in the Northern Hemisphere and the Namib and the Australian deserts in the Southern Hemisphere all lie roughly beneath the same two invisible lines encircling the Earth.

Because of the extremely high air pressure, air just builds up and builds up and builds up here, fed by the Hadley cell pushing its air northward, and forms into a sloping tower of air. Winds that come up against this barrier manage to get partway up its slope before sliding right back down again, like your uncle trying to navigate the icy driveway of your parents' house after too many highballs, except with less cursing.

Why, you may be asking yourself, doesn't the air simply continue its travels north? Especially if there's this huge swell of air that wants to get anyplace with a lower pressure. How does a big pile of air like this build up in the first place? It doesn't make any sense!

Turns out we were just getting to that, so calm down.

The Ferrel Cell

Just north of the horse latitudes, also feeding air into its high-pressure system, is another big puffy inner tube of air rolling over itself like a conveyor belt, similar to the Hadley cell, except that it moves in the exact opposite direction, moving southward at the top and northward at the bottom. Both cells, working together, keep the horse latitudes stuffed full with air. So, once you get into that dull, high-pressure latitude, you're just going to have to hope and pray for the day when you can push your way back into one of the rolling systems on either side. Once you manage it, though, should you fall into the **Ferrel cell,** you may wonder why you were so eager to leave all that placidity.

This cell, named for William Ferrel, nineteenth-century scientist and popularizer of the cowbell, tumbles its supply of air around in a verticalish circle from around 60° latitude to around 30° latitude. In the same way that the Hadley cell is responsible for the easterly trade winds, the Ferrel cell is responsible for the **prevailing westerlies,** which, as the name might suggest, move in a westerly direction (which is to say eastward). Although the winds here do kind of move in a somewhat westerlyish direction, more or less, they're much less predictable than the easterlies, subject to all sorts of chaotic peculiarities. For sailors who had previously used the sure and steady trade winds to guide their ships from Europe to the Americas, jumping into these westerlies was pretty much their only option. And if they could have found a different option, they surely would have chosen it.

High- and low-pressure systems roll about these latitudes, causing severe changes in temperature, surface pressure and precipitation, making it impossible to guess how Weather will behave from week to week. Much of North America and Europe lies within this cell, and this is why the weather is so uniformly uniform here. Storms bounce around like numbered Ping-Pong balls in a local TV station's lottery machine. Weather can come from any direction, bringing practically anything at any time of year. Snow in April? Okay. A heat wave in November? Sure. A nice, calm, sunny day in March. Seems unlikely, but all right.

The reason for all this craziness is that the Ferrel cell isn't its own self-sustaining cell. In fact it's kind of not an actual cell at all, though it functions as one well enough for our purposes. It's actually what's called a **thermally indirect cell,** with the air rising at the area of coldest temperature and sinking once again at the area of the warmest temperature. If you think that doesn't make any sense, you're right. But it's what happens, and like so many other things in the world that don't make any sense (the origin of life, dark energy, the electoral college), you're just going to have to accept it.

In this cell the air movement isn't being pushed along by the natural movements of heat and air pressure. Instead it moves in direct response to the Hadley cell to the south of it and another cell that lies to the north of it, born of the cold low surface pressures of the polar regions. And it's because of this bizarre heat and pressure dynamic that it's such an unruly area in which Weather may roam free and lawless.

The Polar Cell

Once you've spent your share of time getting pushed around like a pinball in the Ferrel cell, eventually, if you're lucky, you'll get knocked into the path of some warm air heading north and run right into the **polar front,** or **subpolar low,** a low-pressure stormy band around which the air on either side can be markedly different. It occurs at around 60° latitude but can move as far south as the subtropical latitudes in cold winter months.

Just past this polar front, though, you will find the final cell, the **polar cell,** named for . . . well, it's kind of obvious. Once inside this cell, you and the air around you will make your way upward into the scenic vistas of the upper troposphere, where you will then flow pleasantly northward toward the northernmost place on Earth: Santa's Workshop and Five-Star Luxury Hotel.

But remember our old friend Coriolis. While you move northward, you'll also be turning right, taking a westerly direction in the upper air. And you're also being cooled by the chilly polar temperature, so just as you're reaching Santa's place, planning on some nice spice cookies and a hot mud scrub, you come down— still turning right, just as you did in the Hadley cell— and begin going back down south and to the west. Santa's place was so close just a minute ago; you could smell the gingerbread and reindeer droppings. But now it's vanishing behind a mist of powdery snow, as the air around you rushes in to fill the void left by the air at the surface level that's moving down south toward the low,

low pressure of the polar front. This south and west-ward movement of air en masse creates another band of surface easterlies, similar to those in the Hadley cell.

When you head south enough to reach the polar front, you'll once again be lifted upward and then north toward Santa's, but again you won't quite make it. You'll just have to content yourself with a sad wave to Rudolph as you make another trip. One day you'll break free of this cell and back into the Ferrel and Hadley cells. Maybe you'll get far enough south to do the exact same trip (but in the opposite direction) in the Southern Hemisphere. (Except down there, instead of Santa's Workshop, you'll find Biggie and Tupac's Super Secret Recording Studio.)

If you're going to travel with air, be prepared. Air never stops moving. Never, ever, ever, never. It's constantly circulating around the globe, like Phish.

Like Big Invisible Fettuccine in the Sky but Faster

Of course there is another option for you besides blowing back and forth around the planet in never-ending figure eights. But it's not for the faint of heart; it's some serious wind experience. One might even call it Xtreme wind if one were a hackneyed marketing executive. But it's probably not for you. You're not interested in super-turbo-charged winds whipping across the lower stratosphere, high above the rest of the world, like the venomous tail of some furious aerodemon, leaving monstrous storms spinning like tops in its wake, are you? No, I didn't think so. Maybe I'll just save the **jet streams** for some other book. Maybe one about ice fishing. (Oh, those ice fishermen, they're crazy! I could tell you some stories.)

So, moving on . . .

What's that? You don't want me to save the jet streams for the ice fishermen? You want me to explain them here? I don't know. They're pretty wicked. I mean, we're talking 300 mph winds. That's stronger than a hurricane and almost as strong as a tornado. All right. But remember, it was your idea.

As you're floating through the various air cells, next time you come upon one of the barriers (the subpolar low or subtropical high), try not to move on into one of its adjoining cells. Instead, when you rise up to the tropopause, keep your eyes open for a thin band of high-velocity air—only a few hundred miles wide and a few miles thick—like a long, rapidly moving piece of invisible fettuccine with no beginning and no end, moving westerly, wrapping all the way around the planet. Even though you can't see *it*, you'll almost certainly be able to see its effects. It should be pulling wispy bits of high-altitude clouds along its path.

Once you find it, jump in. And don't say you weren't warned.

You'll be able to find a jet stream at tropopause level in both areas where the cells meet each other below. Where the Hadley and Ferrel cells meet, you'll find the **subtropical jet stream** about eight miles up, and where the Ferrel and polar cells meet, you'll find the **polar jet stream** about six miles up. (You'll find two identical jet streams at similar positions in the Southern Hemisphere, but moving easterly.)

These jet streams form as the result of the difference in temperature and pressure between the two cells.

At the altitude where jet streams form, the difference between those two factors is at its most pronounced. The pressure gradient force is really strong, so the air attempts to move very quickly from the high-pressure area to the low-pressure area, but because of the Earth's spin and the lack of friction up there, the air instead goes flying off to the east.

The polar jet stream moves somewhat faster than the subtropical jet stream, and since you were interested in a ride, that's the one with which you're riding. Now, the first thing you may notice as you find yourself thrust alongside the wind—aside from the incredibly fast speed at which you're moving—is that you're not really moving in a simple circle around the globe. No, you're whipped about north and south along the latitudinal lines of the Earth like you're on the most nauseating amusement ride ever. You go south for a couple hundred miles and—*bam!*—you're heading north again. The jet streams are incredibly erratic and unpredictable. Sometimes one jet stream will split off into two smaller, less powerful substreams, and you'll ride along the southern branch for a while, kind of getting used to the slightly calmer going, and then—*whoosh*—you've reconnected with the northern branch, and it's all crazy again. Occasionally, if temperatures are cold enough to bring the polar front sufficiently far south, you may even find yourself colliding and merging with the subtropical jet stream. (Pray that doesn't happen.)

Because of the wavering path they take, the jet streams are responsible for moving a lot of the Earth's warmth back and forth around the globe. While on its way south, the jet stream carries cool air toward the equator, and as it curves back up, it brings warm equatorial air poleward. In doing so, it also carries a lot of atmospheric debris (ash, pollutants, old T-shirt particles, aerosol molecules) to whole other parts of the planet, ensuring that humans all share somewhat equally in one another's filth.

These jet streams also contribute greatly to the chaotic mess in which Weather revels in the mid-latitudes of the Ferrel cell. But we'll see more of that in the next chapter on localized weather systems and precipitation.

Wind Through a Zoom Lens

Now, we're going to need to get you out of this jet stream, so let's say it takes a quick left turn and you get knocked loose. That feel any better? Good. Now you're several miles above the Earth's surface and dropping fast. You just experienced firsthand how air moves around the planet on the **global** or **planetary scale.** We got that out of the way. But that's only one of the four **scales of motion** for air circulation. And it's the largest, so everything from here on out should be a piece of key lime pie.

Anyway, you're hurtling toward the ground, approaching terminal velocity, so now is the perfect time to take a look down and pay attention to what you're seeing. It probably somewhat resembles the weather map from your local TV news, except without the smiling sunbeams. This is the **synoptic scale.** (*Synoptic* means "to look at altogether." Essentially, it's an overall view.) This can encompass areas ranging from a couple hundred miles of surface area (as you're experiencing) or a couple thousand miles. I don't know if you can tell, but you may be interested to know that there's a large wall of cold air moving in on that city beneath you, coming in from the north. And because the city is currently covered by a mass of warm air moving southwesterly, it could mean a storm's a-brewin'.

The air patterns that you're seeing from this vantage could take days or even weeks to rejuvenate themselves fully. This is five-day-forecast weather happening here. You may think it's meandering, but it's actually moving just as fast as air ever moves. It's essentially the same stuff you were floating through in the Ferrel cell. Think of it as if you were watching a Boeing 737 fly across the sky from the ground. It barely seems to be moving from where you're standing, while in reality it's going . . . I don't know—really fast. (This isn't a book about jets, so give me a break.)

As you continue to fall—the city rising larger and larger beneath you, details of its buildings and cars coming into view—you're leaving the synoptic scale and entering the **mesoscale.** (If you still remember what we learned in Chapter 2, you've probably already guessed that this means "middle scale.") This concerns ranges from about a mile to about a hundred miles of surface area. See that house over there? The one with the fire trucks in front of it? Notice how the smoke that rises from its roof is blowing easterly for a mile or so toward the synagogue before curving around as it gets higher and going westerly? That kind of air pattern is typical of the mesoscale. The air movements you experience here are responsible for local winds, such as the kind that blow charred wood particles around the city or whip around corners to knock newspaper pages from your grasp, as well as thunderstorms and tornadoes. At this scale, Weather's work may last less than a day or even less than an hour. In forty-five minutes that smoke may be blowing in a completely different direction.

You're coming down fast. You're about to hit the ground. You can now clearly see the faces of the firefighters, distraught homeowners, and curious neighbors who have craned their heads upward to gawk at you as you fall. You're entering the **microscale.** (*Micro* obviously means "small.") Notice how the smoke is billowing up from the house in little curlicues? Or how family photos rescued from the house are dancing around on the front lawn? Or how bits of grass and dirt spiral around you in the air when you slam hard into the front lawn? That's how air behaves in the microscale. Here little eddies and pockets of air—no larger than a few feet—move about by their own chaotic rules, irrespective of what's happening in any of the larger scales, formed through small differences in air pressure or in response to objects in their path. This is the air that blows leaves from trees, kicks up sand, and sends discarded plastic bags spinning in breathtakingly beautiful spirals of existential ennui. Any function of air movement at this scale may last no longer than a few minutes or so.

Have you got all that? You have a better understanding of how wind works now? Think that a certain editor for a certain publishing house that will remain nameless will be happier with this chapter addendum? Good.

Okay. Pick yourself up, because we've got a lot to accomplish in the next chapter. That cold air mass is coming in from the north, and it doesn't look like it had the best of intentions for that southwesterly warm air mass. And here you don't even yet fully understand how it's moving in or what kind of wet chaos it's bound to cause. (Too bad it couldn't have gotten here in time to save the house, huh? Oh, well.)

CHAPTER 4

MASS ON MASS ACTION!—HOW BATTLES BETWEEN AIR MASSES MEANS WE ALL LOSE

Up till this point we've only yet watched as Weather aligned its soldiers along the grassy hills of the battlefield. The naive and ill-informed might be moved to assume that Weather's not such a bad guy after all. *It's actually rather fascinating,* you might be saying to yourself. *I think it's important to have an appreciation for all functions of the Earth, Weather included.*

But that's where you'd be wrong. Weather is *not* not such a bad guy. He is—or, I mean, *it*—it is a terrible guy. It has only the worst of intentions for you. Sure, from way up high in the atmosphere, where you can watch its winds drift in somersaults up and down the globe, or in the grass of a shady front lawn, where you can watch leaves spin in picturesque eddies, its maneuvers may seem harmless. Charming, even. But need you be reminded of George Lucas? Everybody thought he wasn't such a bad guy when he released the original *Star Wars* and *Indiana Jones* trilogies. And just think of the ceaseless human suffering his movies have caused since. I mean, just *The Phantom Menace* alone! The mind reels at the untold agony. (So every time you read the word *Weather,* you think *George Lucas.* That's the same mnemonic they use in meteorology school, I'm told.)

Now, to understand fully the gravity of the situation before us, we'll need to head back up a few miles into the troposphere, to the synoptic scale, to get another look at those air masses encroaching on the city.

Masses of Air Masses

Remember how, in the last chapter, we observed a mass of cold air moving northerly and a mass of warm air moving southwesterly? Well, those aren't large pockets of air. Those are armies. Battalions of air molecules preparing for battle. And where do you think the battlefield is? Right above that city.

You see, each of those air masses has very different properties, because each was formed in a very different climate, in either Hadley or polar cells. They amass their strength slowly over wide, flat areas of land or sea with high atmospheric pressure and surface winds light enough to allow them to grow in peace, taking on the properties of the land over which they wait, wait, wait . . . wait for strong upper-atmosphere winds—the jet streams—to swoop down and carry them—as huge high-pressure pockets of air several thousand square miles in diameter—into the mid-latitudes, the Ferrel cell.

The temperatures of these latitudes are too murky and inconsistent to form their own air masses, the winds too unpredictable. But this is an ideal location to come for a good fight, to dance around the continents Jerome Robbins–style and rumble. Throughout one year you can find countless air masses like these rolling about through North America, Europe, and Asia, looking to do

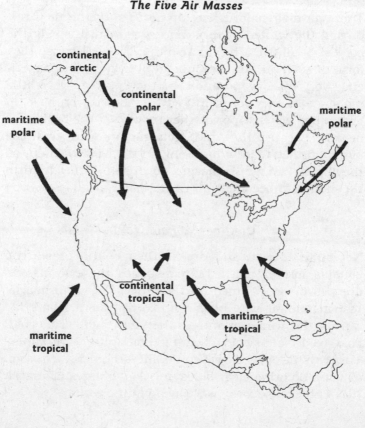

The Five Air Masses

some ass kicking. But they all fall into one of five categories, broken down by the type of surface over which they were formed—**continental (c)** if over land and **maritime (m)** if over water—and the climate in which they were formed: **arctic (A), polar (P),** and **tropical (T).** The combinations create five different types of air mass, each with its own properties, which, coincidentally, are listed below.

Continental Arctic (cA)

These air masses form way up north of the Arctic Circle during the winter months and are unsurprisingly the coldest of all five. They're also the driest. Because they form in such a cold environment over land, they experience very little evaporation, so they pick up very little moisture as they form. In fact they don't even pick up very much water vapor as they move over the cold polar oceans. Because the Arctic gets slightly warmer in the summer, they tend to form only in the winter. Often, as they meet warmer air masses down south, they'll warm up enough to become continental polar air masses.

Continental Polar (cP)

Not quite as cold and dry as their continental arctic cousins, forming a little bit more south, these air masses are still pretty damned cold and dry. They're primarily responsible for why people in North America hate life during the winter months. They are cold, cold, cold, cold, cold. Not as cold as continental arctic? Who cares? They're still cold. However, in the summer their cool, dry airs do bring clear skies and cool breezes. But still, that's no excuse for what they do in the winter.

Maritime Polar (mP)

Bringing cloudy, damp cold that penetrates your clothes and skin and seeps into your bones, these air masses can suck it. They form all year-round above the North Atlantic and Pacific oceans, sucking up ocean water that they will later use to bury your car in snow. They tend to be slightly more temperate and less garbagey than continental polar airs.

Continental Tropical (cT)

The air masses that form over the deserts of the south-western United States and Mexico are hot and bone dry, producing blistering heat that sweeps eastward across the Midwest, causing small dogs to explode on asphalt. Their kind of air needs the summer to make themselves sufficiently horrible.

A particularly volatile example of these are the Santa Ana winds, which form above the desert plateaus of the American West's Great Basin and spill downward through Southern California's mountain ranges, build-ing momentum and heat and losing humidity as they're compressed through the rocky canyons. The desiccated vegetation they leave in their wake make for perfect brush fire conditions, but we'll waste more time on that in Chapter 5.

Maritime Tropical (mT)

You know that old saying "It's not the heat; it's the hu-midity." The only thing more annoying is the kind of air that makes people say it. These air masses are hot and wet, having formed over the southern Atlantic Ocean and the Gulf of Mexico. They slosh up into the southern and eastern United States, knocking on mosquitoes' doors and telling them it's time to come out and feed. They can develop all year-round, but they're particularly strong in the summer. Once they've had their fun in the states, they will often slosh over to the British Isles, where they bring overcast skies, haze, and hill fog.

Air Wars

So, here are all these different air masses moving up and down and around the land masses of the mid-latitudes, all with different air densities, temperatures, and water vapor contents. Here it's not just so simple as saying that high-pressure air wants to get to the low-pressure area. If it were, they'd all just quickly blend together, mix their properties, and we could enjoy autumnal temperatures all year-round. But that's not how Weather plays its game. It pits these different air masses against one another. But why? What happens when one of these air masses runs up against another? What will happen above this city when that northerly cold air meets up with the southwesterly warm air?

The point at which they meet, the point at which the two battalions of air molecules slam into each other—a battle line that can range from less than twenty to more than one hundred miles in length, depending on the size of the air masses and the angle at which they're meeting each other—is called the **frontal surface** or **frontal**

zone, and it's almost always the line marked by the lowest pressure. These generally are not straight vertical dividing lines heading at a ninety-degree angle up into the sky. The air masses are too complex for that. First of all, one of the two air masses is going to be colder, and therefore more dense, than the other, so it's going to want to stay closer to the ground, and the other one will need to climb over it. Second of all, the air particles at the very top of each air mass may be moving about very differently from the ones at the bottom. So what you'll usually get is a curved or diagonal frontal surface.

With the approach of a front, you can expect several different things: sharp temperature changes in a relatively short period of time; a change in the moisture content of the air; a shift in wind direction and a drop in air pressure; and an ever-changing procession of types of clouds. (We'll get more into clouds in the next chapter. Just keep the names tucked away in the back of your head for the time being.)

Cold Fronts

If the cold northwesterly air is stronger than the warm southerly air, what the city will experience is a **cold front.** These are the fastest of the fronts, squabbling their way across the land at something like 9 to 30 mph. They cause short-lived but violent thunderstorms and sharp drops in temperature, sometimes more than fifteen degrees in less than an hour.

As the front approaches, the city will experience a severe drop in air pressure, with increased winds moving *toward* the encroaching front. The sky at this point will already be overcast, as the cold front will usually send thin and wispy upper-atmosphere cirrostratus and cirrus clouds along several miles before it to blot out the sun and make the coming storm sufficiently ominous.

Then, when the cold air mass meets up with the warm air mass, it uses its greater force to wedge itself *under* the warm air, like an offensive line in football, getting down low and forcing the defensive line backward by putting it off its balance. The warm air has nowhere to go but up, and it finds itself sliding up and over the cold air, creating an ovular frontal surface with a steep lower ridge. This steep ridge is the result of friction. While the old air sweeps under the warm air, it's kind of getting caught up on the ground. Try pushing your limp fingers across the surface of your desk. Notice how they bend inward, but not under, as they move, causing a sort of straight line near the desk's surface. That, in a (very loose) sense, is how the cold front moves.

As the warm air gets forced up in the higher atmosphere, its molecules begin to cool. Remember the lapse rate you experienced in Chapter 3, how the air grew colder as you moved higher up into the troposphere. Well, that's pretty much what's happening here. And as the warm air rises, it can't hold as much moisture, so its water vapor starts **condensing** into microscopic **cloud droplets,** about 0.0008 inch in diameter. These collect as cumulus clouds, forming just in advance of the frontal surface (with just a small part of it behind the line), and get pushed forward by the moving cold air.

Within these clouds, a fraction of the cloud droplets will collect themselves into **rain droplets,** acquiring the weight to fall past the winds, causing thunderstorms and heavy winds to pelt the city beneath. Because the frontal surface curves inward, even when the front has passed through the city at the surface level, the higher

Cold Front

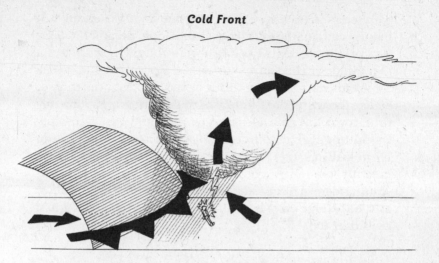

atmosphere above the city is still *behind* the frontal surface. So the clouds and rain will linger, even after the
front has passed them by at the surface, leaving cooler
winds behind to add to the rain. But because the frontal
surface has now moved past the city at ground level, the
winds reverse course and head toward the low-pressure
area as it moves away.

Some of the cloud droplets will make their way up
into the atmosphere *above* the frontal surface, which
may be no more than a mile or so above the surface,
and form into the cirrostratus and cirrus clouds that we
already saw getting pushed along by upper-level winds
ahead of the front.

As the front passes—lower and upper levels alike—
air pressure rises, and the rains end. The air dries out,
leaving clear skies with the exception of a few straggling
low-level cumulus clouds that really just wanna hang
out and chill.

Sometimes, if a cold front is moving fast enough, it may send out a **squall line** a mile or so ahead of it. This is a thin band of heavy rain and strong winds that just can't wait for the front.

Cold fronts usually travel from the north, northwest, or west, with *southerly* or *southwesterly* winds shifting to *westerly* or *northwesterly* winds as it passes. Sometimes, however, cold fronts will move in from the east or northeast—as they do quite often in one of Weather's favorite stomping grounds, New England—with winds shifting from *southwesterly* to *northeasterly*. This is known as a **backdoor cold front.** And yes, I laughed when I first read that too.

Warm Fronts

If the warm air mass has the sufficient chops to over-power the cold air mass, what the city will endure is a **warm front.** Warm fronts generally move slower, last longer, and produce a lot more rain than cold fronts, but in a considerably more laid-back Fonzie kind of way. Less turbulent winds and lighter rain.

A warm front would advance upon the city at about half the speed of the cold front, but it could be hovering over the city well before anyone even knows it's there. Its frontal surface would be shaped like a long and shallow inverted slope. Its highest vertical point could be seven hundred miles in advance of where it touches the ground. That means that for several days the city could continue experiencing cold weather while, five miles overhead, the warm front has already begun to envelop it. At first the only clue to the average person would be the appearance of cirrus clouds in the upper atmosphere above.

The way a warm front works is by climbing up and over the top of a cooler air mass and slowly pushing it from its way. Imagine a puddle of pomegranate juice on your kitchen table. If you were to take your gym membership card and lay it flat on the table and then slide it into the puddle, the pomegranate would just spill up and over it. *But,* if you held the gym membership card at an angle so that the front was just higher than the pomegranate juice and then slid it, the card would cover the juice for a bit and then begin to push it toward the table's edge. The pomegranate juice would get caught under it and be unable to get away, and before long it would be falling onto the linoleum floor below, splattering onto your socks.

Along the frontal surface there's a long area in which warm air is coming in contact with cooler air. Where they meet, clouds form. As it continues moving in, the frontal surface gets closer and closer to ground level, the clouds get thicker and lower, and the sky gets darker. Ice crystals from a cirrostratus cloud limn the

Warm Front

sun in a bright halo. When the front is still more than 350 miles from the city, inhabitants may begin to see falling snow, and the sun is completely blotted out of the sky save for a hazy illumination.

The winds get stronger, rotating clockwise as the low-pressure surface front approaches. Snow turns to sleet and then to a persistent but gentle cold rain. Temperature rises, and then the surface front passes through the city. Its effects are less immediate than the cold front's. The air pressure stops falling, the rain ends, and the warm front goes on its way, leaving a few stratocumulus clouds in the middle atmosphere.

In some situations, if they're sufficiently humid, warm fronts may bring thunderstorms just ahead of their surface fronts, but not usually.

They generally move toward the north or northeast with winds blowing *southerly* or *southeasterly* ahead of the surface front and leaving *southerly* or *southwesterly* winds behind it.

Stationary Fronts

If neither air mass is strong enough to overtake the other one, then what you get is a **stationary front.** Both air masses just stand there trying to look all badass, guns drawn but unable to blow the other one's head off. Things can go on this way for days.

Because there's very little mixing of air between the two masses, the winds tend to blow parallel to the frontal surface but in opposite directions. If you walked from one side of the front to the other, you'd notice a sharp shift in wind direction. Obviously, temperatures will be warmer on the warm side and cooler on the side that's not the warm side. If both air masses are relatively

Stationary Front

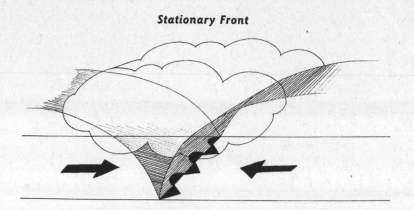

dry, clear or partly cloudy conditions along the frontal zone may occur. If either is more humid than the other, the small amount of mixing that does occur can cause overcast skies or rain lasting for more than a week.

When one of the two fronts gives up the standoff and relents, or one gets a boost of energy, the stationary front will become either a cold front or a warm front. But usually they'll just tire each other out and dissipate.

Occluded Fronts

An **occluded front** (with *occluded* meaning "closed off") is a weird kind of front, responsible for a lot of weird kinds of conditions. As occluded fronts approach the city, they may exhibit all the characteristics of a warm front, but as they pass by, they act as a cold front. They're kind of weird like that.

What they actually are is a cold front catching up with and overtaking a warm front. Both should be moving in somewhat the same direction, and since we've already learned, cold fronts move about twice as fast as

Occluded Front

warm fronts, the cold front runs right into the back of the warm front and lifts it—air mass, front, and all— into the air and pushes it along before it, like when you grab your dog's back legs and push her around like a wheelbarrow, but with less whining. If the occluded front is colder than the air ahead of it, it's called a **cold occlusion.**

If, however, the air up ahead of the occluded front is colder than the air behind it, it won't be able to lift the air ahead, and this is called a **warm occlusion.** As it passes, it will leave conditions similar to those of a warm front.

Okay, so you have warm, moist air and you have cool, dry air, you might be saying. And when they come into contact with each other, they start blowing around all crazy and the moist air that gets pushed up into the sky cools down and that causes the rain that ruins our suede jackets. Yeah, that makes sense, you say.

No, wait. That doesn't make sense, you continue saying. The physics of a storm cloud seem like they'd be a lot more complicated than just water vapor con- densing into cloud droplets as it rises, and those droplets eventually collect as rain droplets, and those droplets fall to the ground as rain, you recall from earlier in the chapter.

Wow, you actually retained that information a lot better than I thought you would on the first read through. Clearly, you are a person hungry for meteorological information, and you will not be denied. I will take that under consideration in the future.

So anyway, since you asked, let's take a closer look at the goings-on inside a storm cloud.

The Economics of a Storm Cloud

When we look up at a cloud floating tranquilly through the sky on a warm summer afternoon, it almost seems as though we could reach up, grab it, tear off a little bit of fluff, and shove it into our mouths, allowing its sugary cloudiness to dissolve on our tongues, while we complain that the line for *The Jiggler* is too long and vomity-smelling.

Many, many hours of research and testing by meteorological scientists have disproved that theory. Unfortunately. Instead science shows us a different kind of cloud, one much less substantial and considerably more muggy.

As we saw earlier, clouds really are nothing more than a whole bunch of water particles that broke free from the ground, got caught up in some winds, and ended up stuck up in the sky (along with some other stray bits of tiny airborne flotsam). To see them from a great distance, you might think that they'd just be bursting full of liquid, ready to fall, but that's really not the case. A typical cotton candy–looking cloud, hundreds of cubic feet in size, probably contains less water than the tank that the developly challenged clown falls into when you hit the bull's-eye with a softball. (You see, that was a callback.) If you've ever walked through a fogbank, then you have a good idea of what a cloud looks like from the inside. In fact fog is just a cloud on the ground. Or maybe a cloud is fog in the sky. Or something. It's immaterial.

Clouds form when the air becomes so infused with water molecules that they all start sticking to one another and eventually collect into a bunch of **cloud droplets.** There are actually several factors working together that make this possible: the percentage of water vapor in a given block of air; the pressure of that air; and the temperature of that. When they're all factored together, you get the **relative humidity** of that block of air.

So we have a block of air relatively well packed with water vapor, the water molecules all bopping around into one another. If we decrease the temperature within the block, the relative humidity will rise, because cold air can't hold as much moisture as warm air, and it's getting closer to the point at which it can't hold any more. If we free up some space by removing some water molecules, the relative humidity will drop to where it was before. Then, if we squeeze them all into a smaller block, the air's relative humidity rises once again. The higher the relative humidity, the greater the chances that water molecules will bind together.

Now, let's assume that a bunch of water molecules bound together and formed a cloud droplet. If the relative humidity remains constant, water molecules are going to be leaving the droplet to go off on their own at the same rate as other molecules are getting stuck to it, so the droplet maintains its size. The cloud droplet is in **equilibrium.**

If, however, the relative humidity drops, the water molecules on that droplet will become more excited and water molecules will detach themselves from the droplet (thus **evaporating**) faster than other molecules are attaching (or **condensing**) themselves onto it. So the droplet shrinks in size. But if the relative humidity rises, molecules will attach themselves faster than others detach themselves, and the droplet grows. Simple enough. But we still have the **curvature effect** to consider.

It takes a lower relative humidity for a plane of flat water to maintain its equilibrium than it does for a free-floating droplet. That's because water molecules won't bind as strongly to a curved water surface as they will to a flat surface. A tiny little cloud droplet has a much more pronounced curve to its surface than a big one does. The smaller it is, the harder time it has maintaining its equilibrium and keeping its molecules, while the bigger droplets not only more easily keep their molecules but also have an easier time collecting new molecules the larger they get. And then, when they get larger, they have an *even easier* time collecting new molecules.

Think of it like money. When you don't have much, it's really hard to hold on to the little you do have. You work and you work, but you end up spending more than you make. But there's that jagoff huge droplet over there with more money than he knows what to do with. Sure, he might be spending more than you on nice clothes and fancy meals, but he's also got investments overseas, so he's pulling in way more than he spends. So he's just getting richer. And the richer he gets, the more he can invest. It's just not fair. The good thing is in a socialist democracy like the one we live in, there are certain rules. Once he grows to about 0.0004 of an inch, the curvature effect becomes moot, and his equilibrium is essentially that of a flat surface of water. So he—and all his other rich jagoff droplet friends—won't just keep growing exponentially.

For a bunch of rich jagoff cloud droplets to keep sucking in more and more and more molecules and eventually grow large enough to become raindrops, the air has to be supersaturated. That means the relative humidity has to be higher than 100 percent. That would be akin to an amazingly strong national economy, in which the country is pulling in more money than it's putting out. This rarely happens. So the rich jagoff cloud droplet's out of luck, right? Too bad. Actually he's not. He's gonna be fine. Because he's got a little trick up his sleeve.

How many rich people can you think of who started out with nothing, dirt poor, wearing old tires for shoes, and ultimately built themselves into billionaires through sheer force of will? They exist, but they're definitely the exception. Most of them started out with some help. Maybe some seed money from Daddy or some insider stock tips. Or, in this case, **cloud condensation nuclei.**

These are the little stray particles of flotsam, such as pollutants and salt particles, that also exist inside clouds that allow water to condense onto them. Many of them are hydroscopic, meaning "attracted to water." When they bind with water, they become solutions, with stronger binds than water normally has. So the curvature effect isn't as effective. When they're small droplets, it's easier for them to retain their molecules, and thus they're given the time to find themselves and grow with less effort, while the droplets without hydroscopic help have to work their way through school. (It would be nice to take two years to explore your inner poet, wouldn't it?)

Anyway, as these rich jagoff cloud droplets continue to grow, they're sucking all the available water molecules from the air, decreasing its relative humidity. That means it's harder for the honest little droplets to keep the few molecules they have, and they eventually evaporate practically out of existence, doomed to live out their lives as water vapor.

Now we have a cloud full of droplets of increasing size. But most of them still aren't big enough to fall as rain. They're too light. The slightest updraft will keep them suspended right where they are. And even if one does manage to fall, as soon as it reaches the lower relative humidity air beneath the cloud, all its molecules will evaporate away long before it hits the ground. The average raindrop is about one hundred times larger than the average cloud droplet. It's heavy enough to break free from the updraft, and it's big enough to keep a good deal of its molecules on the way down. That's the goal.

Most clouds do not produce rain. And even when they do, only a fraction of their moisture escapes that way. It's why places like Texas, with its high temperatures and low relative humidity, can have overcast skies year-round but barely receive enough precipitation to fill a Dixie cup.

What do cloud droplets need to do to achieve their goal anyway? Well, they have two options.

Collision and Coalescence

In this scenario, the rich jagoff cloud droplets, unable to break free of the updraft, fall and then get pushed back up again and fall and then get pushed back up again. While they're going through all that, they're also bumping into a lot of smaller cloud droplets (**collision**). And when they do, they suck those droplets up, like hostile takeovers (**coalescence**). And they get bigger, so they try to fall again, picking up more droplets on their way down; but they're still too light, and the updraft, like the IRS, pushes them back into the cloud and forces them to work within the system.

Now, once the rich jagoff cloud droplets have gone through this process enough times, they've had the opportunity to grow to the sufficient size that they *can* break free. They're sufficiently heavy and full of water molecules that the law of the updraft no longer applies to them. They no longer need to work within the system. So they fall right past the updraft (making sure to suck up a few more droplets on their way down), leave the cloud, and fall to the ground as rain.

Now, you might be thinking that all the smaller honest droplets should just band together, form a giant collective droplet, and do the same thing. That's a great idea. In theory. In reality it rarely works that way. Smaller cloud droplets have much stronger surface tension than larger ones. When two of them come in contact with each other, they usually can't find a common ground and will usually end up bouncing right away from each other. So unless they get sucked up by one of the larger rich jagoff droplets, they stay right where they are.

Ice Crystal Process

If temperatures within the cloud are cold enough, then the rich jagoff cloud droplet doesn't need to wait around that long. He can hitch a ride on an updraft to the top of the cloud, where it's coldest, and freeze into an ice crystal. (This works best within clouds whose tops extend high up into the atmosphere as the result of the cooler temperatures at higher altitudes.)

For this to occur, the air around the rich jagoff cloud droplet must be below freezing. If it is, it may become an **ice embryo,** a microscopic ice structure. Very often an ice embryo will break apart shortly after forming, though larger ones tend to be stronger and are more likely to remain intact. In air -40°F or cooler, even small droplets can maintain their structure, so really high up in tall cirrus clouds, ones that extend twenty-five thousand feet up or more, you'll find a lot of ice embryos. The embryo acts as a nucleus, collecting other cloud droplets that fuse to rich jagoff cloud droplet's side, as it gets wider but remains flat—to state it simply, water forms flat geometric patterns when it freezes because of its oxygen atom's stronger pull on the shared electrons than the hydrogen's—and eventually becomes an ice crystal.

Once he's done with all that, it's even easier for him to attract more molecules because ice requires a lower relative humidity than water to maintain equilibrium. So he sucks up a bunch of supercooled water particles that were just dying to freeze onto something anyway, he gets heavy, and then he starts to fall.

As he falls, though, he's still colliding with other water droplets beneath him, and, as an ever-increasing ice crystal he's bound to shatter at some point. Did I say shatter? I mean diversify. Yeah, the jagoff ice crystal splits his assets between a number of different ventures; each ice crystal piece picks up stray water molecules, grows in size, diversifies again, and before you know it, there's a whole bunch of ice crystals falling through the cloud. It's like a chain reaction.

They fall through the updraft and, if the temperatures outside the cloud are cold enough, will eventually tumble to the ground as beautiful snow. But that's not how it usually goes down. In most cases the snow will melt as it falls through the warm air, and by the time it reaches the ground, it will have reverted to rain. This is how a lot of the rain that falls in middle to upper latitudes is born. As snow. Created by a rich jagoff cloud droplet.

Keeping all this in mind, what do you see when you look at the frontal surface between two dissimilar air masses?

You see low-pressure areas and cool air mixing with warmer air and stealing its heat. You see masses of warm air being forced upward into higher altitudes with decreased air pressure. All these things cause a drop in relative humidity. And as we've seen, a drop in relative humidity contributes greatly to the creation of rain and snow.

And that, dear reader, is where babies come from.

CHAPTER 5
WEATHER SHOWS ITS TRUE COLORS

Okay, so we already have a general idea of how a drop in temperature, or a rise in relative humidity, can cause raindrops to form in the clouds overhead. But that's just the regular run-of-the-mill rain that may cause some annoyance but rarely causes much serious damage (unless you count the ruination of a collection of Eudora Welty short stories that was idiotically left on the back porch by someone who knows damn well that he did it). When Weather really wants to hit us, it has an entire arsenal of more powerful weapons it can use. We have some idea of how to predict some of them, while others remain complete mysteries. We might as well familiarize ourselves with some of them. That way, even if we can't guess when they're coming, we can at least know how we're going.

Thunderstorms

Thunderstorms are almost certainly the rotten trick Weather pulls out of its bag most often. It pulls them out and rains them down more than fourteen million times every year. Just today, by this time tomorrow, it will have caused approximately forty thousand thunderstorms to occur somewhere on the planet. About eighteen hundred are happening right now. Weather likes its thunderstorms.

These form when warm, humid air finds itself inexplicably trapped beneath cooler, drier air. It doesn't even have to be a whole lot of it either. This can occur through oddities of terrain, unequal surface heating, breezes coming in off a body of water, or two air masses colliding (a storm front). It can be a pocket of air as large as your city block or as large as your car. What matters is that it wants to go up because it doesn't like being stuck down on the bottom. As we've already seen, hot air rises, so up it goes, up into the cooler air above. And the greater the difference in temperature between the warm air below and the cool air above, the faster it goes up and the greater the consequences.

This sets off a sort of chain reaction in the sky. As the warm air rises, its temperature drops, its relative humidity increases, and it forms itself into one single cumulus cloud or a bunch of them, none of which is particularly high. It's just entering the **cumulus** or **growth stage,** the first of three stages. The top of the cloud won't be very well defined at all; it'll just kind of blur itself out of existence. That's because the **updraft** of air, caused by the warm air rising, pushes the cloud droplets up above the cloud where the air is much drier, and the droplets evaporate. But once they evaporate, the surrounding air becomes much more humid, so the next crop of cloud droplets can remain liquid at a slightly higher altitude.

While this is happening, all the water that's condensing inside the cloud doesn't need as much energy to sustain itself as a liquid as it did as a gas, so it releases its excess heat, and the inside of the cloud gets warmer. And this cycle just keeps perpetuating itself. It keeps sucking up more and more air, pushing its cloud particles a little higher up each time. In a few minutes' time the puffy little cumulus cloud has formed itself into a gigantic pillar of a cloud—a cumulonimbus cloud, also known as a **thunderhead.**

At this point, get ready, because it's about to enter into its next stage, the **mature** or **get-the-hell-inside-the-house stage.** If you see all this happening before your eyes, right now would probably be a good time to get the hell inside the house. Because two very important things are happening.

First of all, the cloud droplets are getting heavier and heavier, and they're just about to get heavy enough to break through the strong updraft of the storm. And because the updraft is so strong, that means they've probably had to build themselves up significantly in mass to reach that breakthrough weight. So, when they do come down, they're coming down hard.

Also, the moist air of the cloud is sucking in the drier air around it, a process called **entrainment.** This dry air messes up the cloud's relative humidity, causing it to drop. And that causes some of the cloud droplets to evaporate. But if they're gonna evaporate, they're gonna need to suck in some of the surrounding air's heat to use as energy to re-form as a gas. Consequently the air inside the cloud gets colder. Now the cold, heavier air wants to go back down, causing a massive rush of air toward the surface called a **downdraft.** And now the big, fat raindrops that are falling are getting an extra push from the downward-moving air. And it all comes right down on top of you, somewhere between your house and your car (all the doors of which are locked, by the way).

By this point you're soaked, you forget which pocket you put your keys in, the top of the cumulonimbus cloud (or clouds) is reaching up to the tropopause— spreading out across the sky like whipped cream getting squashed against a ceiling—and flashes of lightning and crashes of thunder are putting you in a very irritable mood. (We'll get to those two more in a bit.) Meanwhile the wind descending out of the storm is hitting the ground and pushing out quickly in any direction it can go. Mostly it just spreads out across the surface. As a result, it's pushing warm air from the surface up into the storm and essentially feeding the energy it needs to keep making your evening miserable.

But don't worry, this is going to last for only about a half hour or so before the storm enters its final stage, the **dissipation stage.** You may notice that the air is getting a lot cooler around you. That's because all the hot air is getting pushed away. Eventually enough of it will get pushed away that the storm won't be able to feed itself any longer, and it'll start to fade. The downdrafts inside the cloud are getting stronger, there's less hot air getting sucked up into it, cloud droplets cease to form, and the cumulonimbus cloud starts to wither, bottom up, until all that's left is a cirrus cloud where the top of the cumulonimbus once stood. And then it's all over; the whole thing lasted less than an hour, and you've got to go back inside and get changed.

Unless . . .

If the difference in wind direction in higher altitudes is significant enough, the thunderstorm can last quite a bit longer and build itself into a more powerful system. You see, if the winds moving in the lower atmosphere and the winds moving in the higher atmosphere are going in the same direction, the rain and the downdrafts will keep the updrafts in check somewhat, and the storm will play itself out much like I explained above. But if the lower and upper winds are moving in different directions, the cool downdraft and the rain will actually leave the storm in a different direction from the one in which the warm updraft enters. So that allows the storm to feed itself warm air longer, and there's no cool air bumping up against it to even it out.

If this happens, things can get pretty bad. How bad? Like grapefruit-sized hail landing on your head bad. Or like tornadoes picking up your car and dropping it off in another neighborhood bad. (We'll get to both those phenomena in a moment.)

So the moral of the story is: Thunderstorms are not pleasant.

There are two little awesomely terrifying thunderstorm-related cards that Weather keeps up its sleeve and pulls out on the odd day when it's feeling especially precocious. Those would be **cloudbursts** and **microbursts.** Neither occurs very often, but when they do occur, the effects can be devastating.

Cloudbursts

In a cloudburst-type situation, you're walking along the streets and everything seems normal, fine as foie gras. Maybe there's some clouds in the sky. Oh, that one cloud directly above you looks a little concerning. It certainly is forming rather quickly. It certainly is getting . . . Run!

When the rain starts falling, it's like someone reached up and unzipped the bottom of the cloud and just let all the rain fall out at once. It's a deluge. Like someone turned on a hose up there. The kind of atmospheric assault that turns atheists into theists and theists into atheists.

Really they're just extremely intense and condensed thunderstorms. They tend to form over areas with a lot of sustained and above-average surface heating for the thunderclouds to suck in as updrafts. If the cloud can keep enough warm air coming in and flowing upward, the rain droplets can get as big as they like, but they're not going anywhere. The updraft holds it all up there like the neighborhood bully keeping the raindrops from getting to their drama club meeting. But even the neighborhood bully is going to get bored eventually and take off. When the warm air that feeds the storm gets diverted elsewhere, the updrafts lose all their muscle, and the raindrops just fall.

If you get caught beneath such a sudden storm, you're almost certain to experience at least one or two inches of rain in less than an hour. But that's nothing. Once it reaches five inches within an hour, it's officially official that it's a cloudburst, according to people who officiate such things. Just to put that into some perspective, that's the amount of precipitation that Wendover, Utah—the driest town in the state—experiences *annually.* You're getting a year's worth of Wendover rain in less than the time it takes to drive from Wendover to the nearest nongodforsaken place, Salt Lake City.

It gets worse. These storms are very small and localized. They don't really move around so much, so if a particularly strong cumulonimbus sets down on top of your house and breaks out the waterworks, relax and take it. That cloud can stay there for hours. Back in 1935 a cloudburst burst from its cloud over Colorado and poured down twenty-four inches of rain in just six hours. That's roughly half of what Rosiclare, Illinois—the *wettest* town in that state—experiences annually.

Needless to say, most civic communities aren't adequately prepared to handle rainfall of such biblical proportions. Why would they be? Does the average American town really need drainage pipes large enough to drive a Vespa through?

Flash floods are a common side effect, with the soil beneath low lying areas quickly becoming oversaturated with water and unable to soak in the quickly descending onslaught of rain. Unable to get soaked up by the ground, fallen water instead accumulates aboveground, while taking in runoff water from higher-altitude areas as well. This is of particular concern in the deserts of the southwestern United States, where the ground is often claylike and less absorbent than in most other areas. Expect extensive property damage and overall wetness with possible drownings.

Canyons can be equally effective death traps. In his excellent book on atmospheric oddities, *Freaks of the*

Storm, in which you might "coincidentally" find a goodly number of the anecdotes from this book also related, climatology professor and editor of *Weatherwise* magazine Randy Cerveny recounts the sorry fate of a herd of European tourists who happened to be hiking through one of Arizona's many slot canyons while he was researching cloudbursts nearby. A cloudburst occurred several miles away from the hikers, but its runoff fed directly into their canyon and by the time it reached them had become an angry wave of muck and water that engulfed them, sweeping them off into oblivion. But at least they got to be in his book.

Microbursts

The fantastic thing about **microbursts** is that they come in two deadly varieties: **wet** and **dry.** If you'd like to spend your last living moments admiring the beauty of the landscape as it screams up toward your airplane window, then you might enjoy having your airline flight fall victim to a dry microburst. On the other hand, maybe you're a water sign. Maybe you're more comfortable hurtling toward a soaking, sodden grave, something that puts you in the mind-set of returning to the womb. If that's the case, then you might prefer perishing along with a planeful of screaming passengers in a wet microburst. With luck, you'll get the one you want. Historically, when microbursts do occur, no one, including the pilots, had any idea of what kind of death trap they were flying into. Or that they were even flying into a death trap at all. So you can choose your favorite, but all you can do is hope that you'll get your wish. Keep reaching for that dream.

If you do get your wish, what you'll be experiencing is a severe downburst of wind, contained to an area of less than three miles in diameter, emanating from a cloud that may or may not look menacing. If you're traveling through a harsh thunderstorm, well, you're probably expecting some rough weather. Hell, even if you're flying through a mild thunderstorm, you'll probably be extra certain to buckle your seat belt tight. But what if you're just flying beneath a simple rain cloud into a storm that's more likely to mess up the jet's fresh coat of wax than cause any serious harm? Will you be expecting trouble then? Probably not. And that's one of the reasons that meteorologists were not even aware of the existence of microbursts until the relatively recent past. Just a few decades ago really. Before that, the damage caused by them on the surface was usually attributed to tornadoes or roving street toughs.

Microbursts form when dry air blows into a preexisting storm cloud, lowering the relative humidity, causing evaporation within the cloud and also lowering temperatures. Then the newly cooled air within the cloud wants to rush down and out as fast as possible. If the air below the cloud is dry enough, a microburst may also form because the rain evaporates before it can reach the ground, creating similar temperature issues at the base of the cloud and resulting in a similar effect. Except drier. This is a dry microburst.

The wind that descends from these microbursts comes down really strongly. Once it hits the ground, it rushes outward in every direction, moving in a straight line (unlike most other types of air movement). The surface winds created by this downburst, whipping along at speeds approaching 150 mph, will knock down trees, tear down houses, and push over elderly ladies. These winds have no sense of decency. Which isn't surprising, considering the sorts of downbursts they come from. Really.

For airplanes flying at low altitudes, such as when they're taking off or coming in for a landing, microbursts are particularly dangerous. When the airplane flies into one, it—inexplicably, as far as an inexperienced pilot knows—begins to rise into higher altitudes. That's because it's encountering the swift winds that're getting pushed outward from the force of the downburst. So naturally the pilot will attempt to compensate by dipping the airplane's nose downward. That's his first and last mistake. Because as soon as the airplane makes it through these gusts, it hits the meat of the downburst: lots and lots of air coming down hard and fast. So the airplane gets pushed downward, with its nose already pointing that way. After that, because of the low altitude and Weather's sick sense of humor, there's usually not enough time to make things right, and the airplane goes down. Sometimes right on top of a small, adorable puppy dog that was just sitting there, not hurting anyone.

Many a plane went out like that. Like Pan Am 759, which in 1982 fell victim to a microburst over Louisiana and ended up with an unfortunate 153 fatalities and no survivors. As well as Delta Flight 191, which in 1985 went down over Texas with 135 fatalities and 29 survivors. But modern radar equipment and something called a low-level wind shear alert system (LLWSAS), essentially a highly sophisticated network of anemometers, have been able to warn pilots far enough in advance of such conditions that they can avoid them. And pilots are better trained in what not to do if they encounter one without warning.

Rain That Should Never Be

Unfortunately cloudbursts and microbursts aren't the strangest type of precipitation you're likely to encounter as you delve further into peculiarities of the atmosphere. Far from it. In addition to an oddly tinted rainfall here and there, such as the ominous yellow rain that fell over Munich, Germany, in March 1886, convincing the locals that Satan himself was hurling down sulfuric rain as a sign of the Apocalypse by year's end (see more on why this happens in the section on snow), a great number of disturbing things have fallen from the sky, with no easily digestible explanation, over the years.

Take, for example, the fish that has fallen like a summer shower, not once, not twice, but a gazillion times. There are reports of ichthycipitation in India in 1830, South Dakota in 1886, Coopers Plain, Australia, in 1906, Newcastle, South Africa, in 1909. And on and on and on. Please note that these aren't a few stray mackerel via catapult by someone with a strange sense of humor. We're talking genuine fish falls. In some cases, living fish dropping from the clouds and puddling up on the ground. (Don't mean to harp on this, but Randy Cerveny's book is worth its price for this chapter alone.)

The most likely, though not proven, theory is that sea-roving tornadoes called waterspouts (more information coming up) will occasionally touch down over a body of water, suck as many fish as it can manage up into its funnel, and retreat back into the cloud above it with the fish. Then, at some opportune moment, the cloud's updraft will diminish, and the fish will fall to the ground just as raindrops do. Amazing.

And it's not only fish. This same insanity has also been known to happen with nuts, seeds, coins, snails, snakes, maggots, and—of course the all-time classic—frogs. And you probably thought that scene at the end of *Magnolia* was too ridiculous, didn't you? To quote the little boy character from the movie, "This happens. This is something that happens."

Lest you find yourself suspicious that these claims are coming from the same types of people who would have you believe that getting anally probed by an alien is something you get used to eventually, in 1969 in Buckinghamshire, England, a respected and sober-minded columnist for a conservative London paper recounted her experience while traveling of witnessing thousands of frogs fall from the empty sky onto her car and the road before her. As with the fish and all the other odd forms of unlikely precipitation, the most likely explanation is a twister, waterspout, or some other type of strong wind.

Weather has an incredibly sick sense of humor.

Lightning and Thunder

By the time you've finished reading this sentence, lightning has struck the surface of the Earth about five hundred times. Go back and read it again. Now it's one thousand. (But don't go back and read it too many times; the Earth is only so big, and eventually it's gonna get around to where you're sitting.)

But in all honesty, lightning isn't really anything to be overconcerned about. It's just a massively powerful exchange of electricity that occurs randomly between the surface and thunderclouds and heats the air around it to 54,000°F. Sure, modern scientists don't completely "understand" it or its effects or when or where it will occur. So what? And who cares if it kills nearly ten thousand people worldwide every year? Only about a hundred of those people are in America, and most of them live in states like Wyoming or Florida. When was the last time you said, "The only thing this country needs more than more Floridians is more Wyomingians"?

Lightning occurs as the result of differing electrical charges, between different parts of a thundercloud, or between different thunderclouds, or between the bottom of a thundercloud and the surface of the Earth, or between the bottom of a thundercloud and your face. Again, we don't know *exactly* why it occurs, but we do know that it *does* occur, as evidenced by the ten thousand dead people and countless dead animals every year. Of course scientists, being all uppity about science and all, aren't content with simply saying it happens because God hates when we vote Democrat, so they have to create a bunch of *theories* to attempt to *explain* it.

The prevailing theory goes a little something like this . . .

Clouds pick up their electrical charge when hail and graupel (explained in a bit) fall through the high-atmospheric level, where you can find its supercooled water particles and ice crystals. When the supercooled water runs into the hail, it immediately freezes to it. The freezing process releases a small amount of heat (because water doesn't need as much energy to be ice as it does to stay liquid), which keeps the outermost layer of the ice slightly warmer than its interior. Then, when this slightly warmer hail comes into contact with a cooler ice crystal, it transfers a few of its positive ions to the ice crystal. The hail, having lost positive ions, is now negatively charged, and the ice crystal, having gained positive ions, is now positively charged. Then they say it was a pleasure doing business, and they go their separate ways.

Because the hail is heavier, the updraft within the thundercloud doesn't push it up as high, so it lingers in the middle and lower regions of the cloud. The ice crystal, on the other hand, is lighter and flat so the updraft can push it way up to the top of the cloud. So now we have a cloud with different charges throughout its body.

That might be fine for the hailstones and ice crystals, but electricity doesn't care for this arrangement. It wants everything to be neutral, to have no charge. And it also has precious little time to wait, having all sorts of other electricity stuff to tend to. So it bolts as fast as it can between the two regions of the cloud—or between two different clouds—to neutralize the charges and make everything right again. That's why you'll often see storm clouds lit up from the inside or witness lightning flashing sideways across the sky.

This would be all well and good if the clouds would just keep their lightning to themselves, but they don't. The negatively charged bottom of the cloud is close enough to the ground that it's able to attract the positively charged ions out of the surface. The negative charge of the cloud and the positive charge of the surface reach longingly for each other, but they just can't bridge the divide, because the air between them is a lousy conductor of electricity. So they end up reaching harder and harder, the electrical potential building up stronger and stronger on either end, until it reaches about seventy-five thousand volts for every inch it would need to travel, and they just can't stand being apart anymore.

At this point the cloud sends down a **stepped leader,**

actually a series of small electrical discharges that bit by bit charge the air between the two areas—two hundred feet or so at a time, pausing in between discharges for fifty millionths of a second—until it's made its way all the way to a high point of the surface, usually a tree or a house or—in the absence of anything like that—you.

The surface then shoots up a larger, more powerful **return stroke,** following the same path that the stepped leader took on its way down. That's why, even though the worst of the lightning is actually traveling *up* into the cloud, it looks to us as though it were coming *down.* The cloud and the surface will repeat this process several times in the span of a second or two, making the lightning appear to strobe, until both sides are satiated. Then the surface turns on the television and watches something it TiVoed while the cloud rolls over and cries. It's a beautiful process.

Why You Should Try to Not Get Hit by Lightning If at All Possible

The power of a lightning strike varies depending on the length of space it needs to cover, but it's pretty safe to say that if it were to travel 1,210 feet, it would be just enough to power your time-traveling DeLorean in the event that you found yourself all out of plutonium. It's also enough to kill you good and dead. Good and dead. However, it's not nearly as fatal as people think it is.

Oh, yeah, sure, those ten thousand dead people every year would probably disagree, but for every one of those dead people, there're about five or six who live to complain about it for the rest of their lives. (That's a rough estimate; there aren't very good records for lightning survivors.) Lightning doesn't work exactly the way people expect it might. All those urban myths about people being cooked from the inside out just aren't true. Lightning doesn't travel *through* your body as much as it travels *around* it. And it all happens so quickly—not like even a briefly sustained shock from a radio falling into your bathtub—that it is conceivable to survive it.

Even still, you should do your best not to get hit. First of all, it can really put a kink in your day. One minute you're walking across an abandoned field looking for bottle caps (for some reason), and the next you're bathed in an immaculately bright light for a brief and brilliant instant. Then, if you're lucky enough to wake up, you have to spend the rest of the afternoon peeling your shoes from your feet and trying to disembed the ring from your finger.

You also might wake up to find yourself naked or partially clothed in tattered, smoldering rags. The extreme heat of the lightning (five times as hot as the sun, remember) is enough to vaporize your body sweat instantly, creating an explosion of steam that needs to go somewhere, and if your clothes are in the way, it's taking them with it. And really, there's almost nothing worse than getting laughed at by a group of gawking passersby just after such an already traumatic event.

Then you also have to worry about the small amount of electricity that did course through your body. What did it do? You might not know for a long while. Weeks or months. But odds are it's waiting for you. Most survivors describe a vague change in the mental faculties after such an ordeal. Loss of short-term memory, irritability, inability to concentrate, sleep disorders, and seizures are fairly common among that group of fairly uncommon victims. One survivor described a continual feeling of being on the verge of spontaneous combustion.

On the other hand, a small number of people actually report getting *cured* of various ailments after being struck by lightning. No, seriously! For example, in one fairly well-documented case, a man who had previously lost his sight in a car accident had it restored following his brush with lightning. And his failing hearing got all tuned up as a bonus.

But you shouldn't use that as an excuse to run outside holding a metal pole above your head during the next storm. Incidents like that are freak freak occurrences. The alternative is much, much, much more likely.

Tips on Not Getting Hit by Lightning to Maybe Think about Trying

Tip #1: *Try to not walk around the middle of an open field in the advent of a thunderstorm. In fact just stay inside.* This one's kind of easy to guess at, really. Shouldn't need that much explanation. Lightning usually strikes outside, so you don't want to be where the lightning's striking.

Tip #2: *If you cannot avoid being outside in the event of a thunderstorm, see if you can't get inside a car.* In all honesty, this is just a different version of Tip #1.

Tip #3: *Stay away from areas of elevation, lonely trees, flagpoles, and really, really tall people.* The lightning is going to seek the shortest distance between the cloud and the surface. So anything sticking up or more highly raised is a perfect target for lightning.

Tip #4: *Keep low to the ground while actually touching it as little as possible.* You may think you're being smart by lying flat on the ground like a pancake, but you're not. Where do you think all those positive ions are getting pulled from? You want to touch the ground as little as possible so that if lightning does strike, very little of it will need to pass through you.

Tip #5: *Avoid wearing metallic bodysuits.* Metal is a fantastic conductor of electricity; it's a terrible fashion statement. I'm going to level with you here: You look like a douche bag. What do you think you are, a robot?

Tip #6: *Stay away from robots.*

Tip #7: *Do not taunt the lightning.* Cursing God, tempting fate, or trying to impress your friends with such idiotic behavior does nothing more than waste valuable time that could be spent screaming and running for shelter.

Tip #8: *Watch out for general weirdness.* If your hair starts to stand on end or you feel your skin begin to tingle, that could very well be positive energy rising through your body toward the sky. Or it's a ghost. Either way, get the hell out of there!

. . . and the Thunder Rolls

When lightning strikes and it heats up the surrounding air to 54,000°F (just in case you forgot), don't expect that air to just sit around and take it quietly. No. That kind of immense heat change causes the air to expand immediately. And not just expand nicely, like "Hey, other air, move out of the way. I'm coming through." It's like a bomb; it explodes. *Boom!* The explosion of air causes a shock, which in turn becomes a sound wave. And that's all that thunder is. It's rapidly expanding air caused by the heating element of lightning. Without lightning, there is no thunder. Sometimes you might think that you hear thunder up in the sky when lightning's no-where to be seen, a low, rumbling growl of thunder that doesn't sound anything like an explosion. But let me tell you one more time: Thunder never travels without light-ning. And for that matter, lightning never travels without thunder. It's a by-product of an extreme emission of en-ergy. Believe me, you'll understand pretty quick if light-ning ever happens to strike anywhere near you; it's very easy to imagine the sound as an explosion.

So, what is the deal with that low, rumbling growl? Why doesn't that sound like an explosion? Here's the deal: Lightning moves fast. Like really, really fast. Like 300 thousand miles per second fast. That's the speed of light *and* the speed of electricity, which really only makes sense because it is light and it is electricity. So yeah, it moves. By comparison, thunder's practically taking the bus. It has to travel at the speed of sound, which is only like 340 miles per second or something ri-diculous like that. Actually it's kind of embarrassing.

What does that have to do with the low, rumbling growl? I was just getting ready to tell you. Let's say a huge bolt of lightning, coming from a cloud six or seven hundred feet in the sky, strikes your favorite pizza joint on the other side of town. (I can't imagine why you're traveling all the way across town to get a slice of pizza, but you must have your reasons.) The time that elapses between when the stepped leader first ventures out from the bottom of its cloud and when the return stroke answers it is for all intents and purposes practically nil. As far as any of us can tell, it's all occurring instantaneously. You might as well say that Gianni's life's work was destroyed in the blink of an eye. All those months he put off installing a proper lightning rod because he was too busy collecting Mario Lanza records or whatever it is that pizza guys do, and then, just like that, it's gone. Because lightning moves quick.

And not only that, but the light from the lightning moves just as fast. So you might as well say that when you saw that lightning bolt come down on the other side of town, you saw it at the exact time that it was happening. Human brains can't even comprehend speeds like three hundred thousand miles per second. Just think of how long it takes for us to realize we're locking our keys in the car.

As for the thunder, it's just being created in response to the lightning, all along the path of the bolt. So that means it's being created on the splintered roof of Gianni's Pizza Emporium at the same time that it's being created six or seven hundred miles away at the base of the cloud. Once it's created, it goes off in its slothlike way across town toward your ears. That lightning takes five whole seconds to cross one mile. Think about that. Count to five. That's how long it takes for thunder to travel one lousy mile.

Now, Gianni's is about seven miles away from your home. That's why you saw the lightning but then didn't hear the thunder until . . . I don't know . . . What's seven times five? Forty-two? That's why it took forty-two seconds for you to hear the thunder after seeing the lightning. (Wait, that's not right. Note to self: Buy a calculator before turning in the final edit.) And now think of this: That lightning bolt was six or seven hundred miles long, stretching from the cloud to Gianni's. And all the thunder that was created from expanding air around it was created at the same time. Gianni's might be only seven miles away, but all the way up that lightning bolt were points that were farther away. So, the farther away the lightning, the more likely that the thunder is going to reach your ears more gradually. And then you also have to take into account that a lightning bolt isn't really one lightning bolt. It's a series of back-and-forth strikes from the cloud to the surface and back again. Each one of those is creating its own thunderclap. So what's reaching you at your house is actually a bunch of thunder sounds that have had the time and the distance to meld themselves into one low, rumbling growl.

Truth be told, that's just one of the reasons that thunder rolls. It's also affected by the echoes, muffles, and all the acoustical tics it has to endure on its way through the landscape from Gianni's Pizza Emporium to your ear. And also, it makes it more ominous. Weather likes to be ominous.

Red Sprites, Blue Jets, Purple Horseshoes, and Elves

Contrary to what you might think, and to what scientists have thought for most of the past several thousand years, lightning does not only strike downward, which is to say back and forth between the clouds and the Earth, but also strikes upward into the sky, diagonally into the sky, and sideways in all directions at once into the sky.

It appears that there's a lot more going on up there in those thunderclouds than we ever thought or can currently explain.

Pilots for years have been reporting weird and colorful happenings up in the sky, particularly when flying above electrical storms—and even more so after several martinis. However, with science lacking the means to study the atmosphere at that altitude, their claims were mostly ignored until recently, when, in 1989, researchers from the University of Minnesota recorded phenomena that came to be known as **transient luminous events** or **upper-atmospheric lightning,** that any of that stuff was taken seriously. Or seriously enough to warrant a few paragraphs in a book that otherwise concerns itself with more pressing matters of atmospheric science.

Take, for example, **red sprites.** These are red plasmalike flashes that emanate directly upward from the tops of storm clouds—usually at the tropopause—and into the ionosphere. They appear in the sky for just a few milliseconds and then vanish because as we've already determined, electricity is fast. They neither flicker nor last long enough for the human eye to perceive them fully, but real fancylike cameras have determined that they're actually a bunch of balls of electricity, ranging in diameter from thirty to three hundred feet. It's believed that they occur when the upper atmosphere builds up a strong positive electrical charge after a particularly cacophonous lightning strike below, and the excess electrons in the upper part of the cloud strike through electrified air above, reacting with the nitrogen in the air above it.

Similar to red sprites, but actually completely different, are **blue jets.** These are narrow blue cones of electricity that shoot out at angles from cloud tops.

They occur much less frequently and at much lower altitudes—about twenty-five miles above the surface—than red sprites so information on them is scarce. They don't appear to be connected with cloud-to-surface lightning but are usually found, for some reason, shooting out from the types of thunderclouds that produce hail. They probably occur in reaction to strong positive electrical charges building up in the air around the cloud top.

And then we have **emissions of light and very low-frequency perturbations from electromagnetic pulse sources,** more popularly known by its very difficult to justify acronym, **elves.** These are flat, dim red, expanding pulses of energy that expand outward horizontally in the ionosphere directly above thunderstorms. They last just a millisecond, but if you could see them better, they'd somewhat resemble the ring-shaped explosion that comes out of the Death Star in the digitally ruined version of *Star Wars.* They appear often to occur in conjunction with red jets, though they happen first, don't last as long, probably aren't actually related, and probably don't actually occur in conjunction with them at all. That's science for you.

St. Elmo's Fire

Not completely unlike a new horizon underneath a blazing sky or an eagle flying higher and higher is the spectacle of a weird little electrical phenomenon that occurs at night in which ordinary pointy objects that are supposed to glow will begin to glow. It's called **St. Elmo's fire,** in honor of St. Erasmus of Formiae, the patron saint of sailors (who was also known as St. Elmo, presumably by people who didn't much like him), because this bizarre light show could often be seen on the masts of ships.

If you were lucky enough to see it, what you would see would appear to be bright blue or purple fire emanating from the piked tips of objects that are electrically grounded. From what I understand, it can be rather beautiful and mesmerizing. But perplexing. How could something like that happen, and more to the point, how could it happen without burning up the objects from which it's emanating?

The thing is, nothing is actually burning. It's really nothing more than a very low-voltage continual electrical spark called a **glow discharge,** similar to what you see inside your local bar's neon light sign. It's caused by a charge in the air resulting from a thunderstorm or snowstorm. You can see it more easily on pointed objects because they focus the area where a spark can jump into the air, and thus it's more easily visible. Since it needs just the right conditions, it doesn't happen very often. And it doesn't matter if you're a man in motion or not. No pair of wheels is gonna take you where your future's lying to see St. Elmo's fire unless it's meant to be. You'll just have to wait and hope like everybody else.

Ball Lightning

A thousand times stranger than anything you'll find in the upper atmosphere or on the tip of a flagpole is the kind of lightning that often occurs at the surface. And when I say "often," I mean, *not very often,* but more often than *never,* which is how often you'd think something like this would occur. **Ball lightning** is actually a ball of lightning, a sphere of dimly glowing electricity that appears out of nowhere and then just kind of floats, rolls, or bounces around for a while, passing through screen doors and windowpanes and chimney flues, terrifying anyone who sees it, electrocuting anyone who comes into contact with it, and burning holes through various pieces of furniture and clothing it comes up against. Sometimes it simply floats away, never to be seen again. Sometimes it vanishes as mysteriously as it appeared. And sometimes it explodes. Different lightning balls act differently, and they even vary in appearance. They might be white, orange, blue, green, or what have you, and they can range in size from one inch to six feet in diameter.

I know. This is the most unreal-sounding phenomenon that you're likely to encounter and one with virtually no hard evidence of its existence. Just eyewitness accounts. But those accounts number in the thousands, and like ghosts, UFOs, and decent frozen pizza, for the longest time ball lightning was considered nothing more than a myth by the scientific community. However, unlike the cases of ghosts, UFOs, and decent frozen pizza, the scientific community is beginning to come around and accept its veracity. And although no one has yet come up with a good explanation for why or how it occurs, it has been re-created under laboratory conditions. Nikola Tesla, eccentric physicist and inspiration for a bad hair metal band, is believed to have created three-foot balls of lightning completely by accident while attempting to make one of his crazy machines do something crazy. In 1991 a group of Japanese scientists supposedly created it using a machine not dissimilar to a microwave oven. (I never did trust those things.)

Some scientists think that it might be some form of ionized plasma with a small magnetic field holding it together; others, that it's a bunch of nanoparticles set loose from sand and illuminated by oxidation. It's also possible that there's more than one form of ball lightning. When it does occur, which is not often, it tends to be during a thunderstorm, but lightning balls have also been reported to pop out of nowhere on beautiful cloud-free days. So your guess is as good as anybody else's.

Even though they appear so infrequently, word of their antics gets around. Back in 1952 a lightning ball reportedly floated through the open window of an aircraft and harassed the pilot for a bit, burning his hair, eyebrows, and seat belt, before floating into the passenger area and exploding loudly and obnoxiously. In 1791 one supposedly teased a young Italian girl, bouncing around her feet, and then in a very uncouth manner flew right up the poor girl's dress and burned its way back out through her blouse and exploded before her eyes. One notable lightning ball had an admirable sense of irony, having materialized inside the laboratory of early lightning researcher G. W. Richman in 1753 and promptly attacked the curious man, killing him in its sad and hilarious immolation. At least one plague of lightning balls has been known to attack an entire town. In 1853, Mount Desert, Maine, was beset with a gaggle of lightning balls during a blizzard. They came pouring into the townspeople's homes, entering through windows and doors and down chimneys, causing mass mayhem, hysteria, and no small number of injuries. One man was shocked so badly by one that he was left stunned and incapable of speech.

So, yeah. Lightning balls. Weird stuff.

Tornadoes

Here's something to feel patriotic about: Tornadoes can occur practically anyplace in the world (except Switzerland, Finland, and Liechtenstein—all tornado-neutral), but where they really feel at home is America. That's right. They could be spinning around at 300 mph, cutting a swath across Warsaw, Beirut, or Abuja, and they might spread the love around from time to time, give them a little taste of the twist. But when they just wanna kick it old school and disintegrate a church, defeather a few chickens, and carry a pickle jar across county lines and deposit them unharmed on the ground, they know where to go. The Midwest, baby—Tornado Alley, U.S. of A.

That's right. We are the tornado capital *of the world!* The United States hosts approximately a thousand tornado occurrences annually. We lose more babies and schoolhouses and pickup trucks to tornadoes than anyone else. And that's something to take pride in.

You should also be happy, if you don't happen to live in the Midwest, that you don't happen to live in the Midwest. Because they like get about a thousand tornadoes there a year. And that is dangerous and scary. These things are the elite ninja assassins of Weather's evil army. They form mysteriously out of the winds of a particularly strong thunderstorm, they eff crazy amounts of stuff up, they do weird head-trippy things with the few items that they don't utterly annihilate, and then they just vanish. It all usually takes place in a matter of minutes. It's like one minute you're looking around at your beautiful house, thinking about how much you love your wife and your kids and your dog. And then *ten minutes later* you're standing in a pile of toothpicks, a childless widower, and your dog has somehow got its head stuck up inside its own anus. And you have no idea what just happened. It's the equivalent of being zapped to one of the layers of hell, except in hell you don't have to deal with incompetent FEMA workers rooting around through your toothpick pile.

Tornadoes begin their short life span in the **dust-whirl stage,** as airborne swirls of fine dirt spiraling up into the air toward a snub-nosed little funnel reaching out from the base of a cumulonimbus cloud. At this stage they're absolutely precious. Honestly, if you can look at an innocent little dust whirl and see an impending monster . . . Well, you have no heart.

In the **organizing stage,** a tornado's winds start whipping around a lot faster. It acts like the rules don't apply for it and does whatever it wants without thinking of the possible consequences. Its weird sky funnel is getting a lot closer to the ground than you'd probably like.

By the time it's reached the **mature stage,** it's somewhere between three hundred and two thousand feet in diameter. And it begins its short four- or five-mile journey across your backyard and toward the bait shop, a gigantic spinning top with its head inside a thundercloud, as loud as a thousand freight trains, sucking things up and hurling them up into the sky. Sometimes it'll have a few smaller, thinner tornadoes, called suction vortices, spinning around it like cruel sycophants.

Now comes the **shrinking stage,** in which it just gets too tired too quick. You know, it could probably pick up the Lincoln Town Car if it really wanted to, but its suction isn't what it used to be. Its width has noticeably decreased, and mostly it just wants to suck up DVDs about the Civil War.

Then, inevitably, comes the **decay stage.** With some tornadoes, you can't imagine their ever actually getting to this stage, but they always do. The less said about this stage, the better. It's too depressing. By this point it's really thin, and well, let's just say it has a hard time maintaining the liquids it sucks up.

And then it disappears. And then some country singer guy writes a song about your town, and he somehow manages to twist the facts around so that it's a parable about why we need to protect our right to bear arms. And then some girl in Delaware puts it on her MySpace page. That all makes about as much sense as the tornado did, so who cares?

You know, we still don't understand exactly what causes these horrible things. We know that they thrive in the spring and early summer, when warm, moist air is more likely to find itself trapped beneath encroaching cooler, drier air above it, causing strong updrafts. What's most likely happening is that if, as the warmer air moves upward through the cooler air above, it meets a series of varying wind directions leveled in just the right way, it sends the upward-moving air spinning, which causes a stronger vacuum at its base, pulling in more air and increasing the force of its spin until it eventually—if the wind conditions remain like that—forms into a tornado. But there's still a whole lot that we need to learn about tornadoes, which is why we still have a bunch of idiots chasing them around the Midwest in trucks and why we have to suffer through garbagey Bill Paxton movies like *Twister.*

The damage inflicted upon us by tornadoes truly is unquantifiable.

Tornado Weirdness

For every one worthwhile thing we don't understand about tornadoes, there are at least three dozen totally effing bizarre things we do know about them. For example, we know that they have a strange affinity for naked chickens. There are countless stories of tornadoes swooping into a farm, plucking all the feathers off all the chickens, and then swooping away to do other ridiculous things. They also seem to enjoy playing with toy cars, by which I mean real cars. They are forever picking up whole automobiles—sometimes with the passengers still inside—and placing them down again someplace appropriately confounding, such as on top of a building or inside a tree.

There have been stories of cows being swept up by a twister's sucking and hurled miles away to land before perplexed and mortified innocent observers. There have even been many rumors of entire herds of cattle becoming airborne and flying away from their pasture like migrating geese. One particularly meanspirited tornado in South Dakota in 1955 took a nine-year-old girl and her pony hostage as they were attempting to gallop home to safety. The girl's mother watched in amazement as the girl and her pony flew through the air more than about a thousand feet. The girl landed on the ground practically unharmed. (No word on the condition of the pony.)

Another tornado, in Saskatchewan, Canada, in 1923, plucked a baby girl from her carriage before her parents' eyes and left her unharmed in a shack two miles away. An adult Missouri woman, her son, and a friend of the family were lifted up together in 1899 and in time safely dropped back to the ground. The woman remained conscious through the entire ordeal and re-

THE FUJITA SCALE FOR MEASURING HOW BAD A TORNADO SUCKS

SCALE	CATEGORY	WIND SPEED	RELATIVE FREQUENCY	EXPECTATIONS
F0	Weak	40–72 mph	38.9%	Light damage: some broken branches; signs knocked down; T-ball games ruined.
F1		73–112 mph	35.6%	Moderate damage: roof surfaces peeled away; mobile homes knocked over like mobile homes; cars that shouldn't be driving in the middle of a tornado anyway pushed off the road.
F2	Strong	113–157 mph	19.4%	Considerable damage: roofs ripped from shoddily built homes; mobile homes essentially atomized; boxcars overturned; hoboes and troubadours run screaming; big-ass trees pulled right up by their roots; knickknacks turned into deadly projectiles.
F3		158–206 mph	4.9%	Severe damage: expensive homes given brand-new skylights; trains overturned; seven people who still ride in trains run screaming; some deforestation; Humvees picked up and tossed around like Fiats.
F4	Violent	207–260 mph	1.1%	Devastating damage: well-constructed homes deconstructed; cows turned into deadly projectiles; portal to Oz beginning to open.
F5*		261–318 mph	< 0.1%	Incredible horrible monstrous damage: Humvees turned into deadly projectiles; strongly built homes give up and disintegrate themselves with dignity; all logic suspended; entropy reigns supreme; the dead reanimated as cannibalistic zombies and turned into undeadly projectiles.

The scale actually goes all the way up to F12, but since wind speeds above 318 mph are pretty much unheard of, there's no reason to go any higher here. However, expectations in the upper half of the scale include such things as Oklahoma's being picked up and dropped down over Oregon and human souls being sucked directly from the chests of their still-living owners. It also goes lower than F0 into the negatives. At around –F3, tornadoes will actually come into your house and make you a sandwich.

called being spun wildly around and around as the buildings of her town grew smaller and smaller beneath her. At one point she looked around and found a horse flying unhappily beside her. In 1896 a tornado was less interested in carrying away three Missouri women than it was in carrying away their clothes. The women ran screaming through the streets of St. Louis as the twister blew up under their skirts and ripped their clothes away piece by piece until they finally escaped to shelter with some small scrap of clothes and dignity left.

Apparently, it's not just naked chickens that tornadoes like.

Waterspouts

Waterspouts, which greatly resemble tornadoes spinning about across the surface of a body of water, are in fact nothing more than tornadoes spinning across the surface of a body of water. That's really it. That's what they are. Sometimes they're tornadoes that have even formed over land, just the same as any other tornado, and then happened to roam out over the water. Those ones are called **tornadic waterspouts,** and they're just as badass on the water as they were on the land; they just don't have as much to destroy. Which is kind of sad, if you think about it.

The other, more common form of waterspouts are the **nontornadic** or **fair-weather waterspouts,** known for their lack of reliability among tornado friends when life kinda just starts going to hell, as it will, even for tornadoes. They are the much less dangerous of the two, with thinner funnels and wind speeds that rarely exceed F0 levels on the Fujita scale (named for storm researcher Tetsuya Fujita, who developed the scale, possibly, though highly improbably, to impress people at the bar around the corner of his house). The manner in which they form is similar to other tornadoes: beneath a cloud under unstable air conditions—surface air warmed by the water beneath it; cooler, drier air above—and often in tropical waters, though they do make appearances in the cooler waters off the coasts of New England and California from time to time. A lot of times you'll see several of them forming at once, all rolling around on their own.

Although fair-weather waterspouts are of less concern than their land-based cousins, that doesn't mean you should go steering your party boat into them. They're still tornadoes, and you're still on a boat, which isn't exactly the most secure place to be. If faced with the option, avoid them. Besides, there's a small chance that they'll come up onto land anyway, behaving then as a weak regular tornado. At least there, if things go badly, you won't have to swim home. But just hope it doesn't bring its buddies. Five or six land-roving waterspouts in one small area, as were seen on the island of Cyprus in 1969, can get a little hairy.

Hurricanes

You may remember all the way back to The Introduction (we were so young then, weren't we? You've really let yourself go . . .) when we discussed that wonderful piece of American history that will forever make little girls named Katrina feel a vague sense of guilt. (Girls, we know you didn't destroy the entire city of New Orleans on purpose.) Well, *that* is a **hurricane,** and *that* is probably the most destructive, cruel, and unnecessary weapon that Weather has in its big weapons chest. Tornadoes are one thing; they're crazy powerful and dangerous, but they're small. They don't last all that long.

But a hurricane . . . you can see them mother-scratchers from outer space! They're *hundreds* of miles from end to end. They look like the gigantic blurred blades of the world's largest desk fan, but without the protective grating, so anything they land on just gets pulverized. Winds and thunderclouds just spinning and spinning at phenomenal speeds, sending raindrops pelting against your face as you try to shield yourself in vain. Water levels rising and waves beating against your car and your front lawn, while floodwaters pound against your front door. Will you let them in? You might not have a choice.

If tornadoes are the elite ninja assassins of Weather's evil army, then hurricanes must be the marauding battalions of bloodthirsty barbarians. They're not content with killing hundreds of people. They kill *thousands* of people. Sometimes *tens* of *thousands* of people. Sometimes *tens* of *tens* of *thousands* of people.

In November 1970 a hurricane that formed in the warm waters of the Bay of Bengal, off the coast of southern Asia, tore through West Bengal in India and East Pakistan in, well, Pakistan. Hurricanes are measured on a scale of Category 1 to Category 5, with 1 being the most refreshing and 5 being the most deadly and unrefreshing. This was a Category 3. (You might think that's not so bad, but you may recall that Katrina was a Category 3 when it made land in Louisiana. And you know how that turned out.) Winds picked up to 130 mph, but since in that part of the world, people use kilometers instead of miles, it meant the winds reached 205 km/h, which just made matters that much worse. Waves rose *50 miles* into the air. (Actually that's not true. I embellished a bit. It was actually 50 *feet,* but when I wrote that, it didn't seem so impressive. But if you ever saw a 50-foot wave coming at you, you'd pass your lungs through your lower intestines.)

At the time India and Pakistan weren't getting along so well, so even though there was a lot of information coming into the Indian government about the brewing hellstorm, it's not clear why the East Pakistani people didn't seem to have any idea they were about to get demolished. People just weren't prepared, and colossal towers of water swept over them by the thousands, carrying them back out into the bay, where they either died or learned to breathe underwater real quick. Either way, most of them were never heard from again.

All told, when Weather had had its fun and decided to recall its barbarians, somewhere between three hundred thousand and five hundred thousand people were dead. The difference between those two numbers is about the total population of Jersey City, New Jersey. When you can't narrow a death count down any closer than that, you're talking about a lot of dead people. To put it into even better perspective—in case you're having a difficult time wrapping your head around nearly a half million people dying all at once—more than half the local population had died. That's what "*tens of tens of thousands* of people" means. Sure, I could have said hundreds of thousands of people to begin with, but it was more dramatic to do it this way.

Hurricanes don't just form out of nowhere; they're not just really big storms that occur randomly. They're complex and precise Weather machines. Luckily for us, there are all kinds of things that can go wrong in the formation of a hurricane. For every hurricane that Weather could conceivably create, only a fraction of them turn out. Thank God.

First of all, a hurricane needs to feed off heat energy. And the only place it can get a strong enough supply of heat is in the tropics—often off the coast of Mexico or in the North Pacific between July and November—where it finds water temperatures of 80°F or higher over a large distance; that's why you don't see the coast of Nova Scotia getting barraged. It also needs light winds, high humidity, and a strong Coriolis force to get its winds a-spinnin'. Hurricanes don't occur close to the equator because the Coriolis effect is practically nothing down there. Most of them form between the latitudes of 10° and 20°.

A hurricane will begin forming when a couple different thunderstorms converge on each other over a somewhat large area of warm water. Many of these storms may be dissipating hurricanes themselves or hurricanes that never quite made it, often, in the Atlantic, originating as the dust storms of Africa. If the surface winds are spinning properly, such that, in the Northern Hemisphere the storms are sent turning in a counterclockwise (or **cyclonic**) direction (in the Southern Hemisphere the opposite would need to be true, with winds blowing clockwise, or **anticyclonic**), the storms form, Voltron-like, into a **tropical disturbance** (or a **tropical wave**). It's not a hurricane yet, but it's on its way to becoming one.

While the tropical disturbance is spinning, it's sucking up the warm water beneath it, gaining energy to spin faster and stronger. It's also generating its own energy by condensing water vapor into rain particles and keeping the residue heat from that process for itself. The disturbance keeps getting more powerful and volatile, and cumulonimbus clouds and thunderstorms develop in its curving movements, and the air pressure toward its top rises. Consequently the higher-pressure air begins to get flung outward, toward the area of lower pressure on the outside of the developing storm, which in turn causes a low-pressure area below. Surface winds rush inward, spinning counterclockwise with the storm winds, pulling all the winds in tighter.

Remember that Sit 'n Spin toy you had when you were a kid and how you would just turn yourself around that little steering wheel in the middle until you couldn't differentiate the color blue from the smell of chocolate? (What a terrible gift to give to a child.) Remember how when you pulled your body in really tight to the steering wheel, you could make yourself spin even faster and get even sicker? Well, that's pretty much the exact same thing that's happening here. As the winds cluster tighter, they spin faster. They still have the same energy, but they don't have as far to go around the center, so they just move around it more quickly. This increases the power of the disturbance, and as the disturbance gets more powerful, it sucks up more energy. It's kind of a cyclonic cycle. Before long the winds are going 25 to 40 mph, and the tropical disturbance is upgraded to a **tropical depression,** named for both the decreased air pressure beneath it and the general mood of any islanders who happen to see it forming.

This just keeps on going. It gets stronger and the surface air pressure drops and it converges which makes it faster which makes it stronger which makes the air pressure drop which makes it converge which makes it faster which makes it stronger which . . . well, it goes like that for a while, until the winds pick up to 74 mph. At that point it's a full-on **Category 1 hurricane.** And what you may notice is that it now has a distinct doughnut hole in its center. That is called the **eye.** Right there in that eye, everything is calm. There are very few winds and no clouds in the sky. It's ironic, because completely encircling the eye is the area of the strongest winds and most intense storms, the **eye wall.** The winds of the eye wall are spinning so fast, trying so hard to get into that eye, but because they're spinning so fast and trying so hard, they can never quite make it. (It's called **centrifugal force**—not to be confused with **centripetal force,** which acts in the exact opposite direction, *toward* the axis of rotation—and if you don't like it, you can talk to Isaac Newton about it.)

While all this is happening, the storm is now like some horrible, evil gyroscope, spinning aimlessly across the surface of the ocean. If, in its travels, it continues to encounter warm water and conditions that keep it spinning, it will continue to gain energy and momentum. It's already increased many times in size, but it can get really big, possibly to several hundred miles in diameter (or half of several hundred feet in radius).

There are some things that will bust up a hurricane. Strong winds, particularly if higher winds are moving in different directions at different altitudes. This tends to mess up the hurricane's groove, and then its winds lose focus and go off in different directions. A loss of heat will starve it. So if the hurricane spins too far north into the cooler high-latitude waters, it'll spin itself right out of a good thing. Same if it goes too far south. The Coriolis effect peters out closer to the equator, and it needs that Coriolis effect to send its air spinning off clockwise from its top, so it can maintain its reactive cyclonic movement at its base. Otherwise, again, the winds lose focus and dissipate.

Another thing that will definitely bust up a good hurricane is hitting land. When it loses its supply of warm water, its clock starts winding down. But you know, killing a hurricane with landfall is not the way to go. For the people whose homes are on the land that's stopping the hurricane's spin, it's a little like jumping on a grenade.

The first thing you can expect is increased wave height at the shore; it might come as a harbinger a few days in advance, creating swells thirty to forty feet high. Enjoy the surfing now, because that surfboard is going to get rammed through a wall soon enough. As the hurricane approaches, the air pressure will drop, which will cause the sea level to actually rise. You might arrive at the beach to find it not there, with waves lapping your feet in the parking lot. Then comes the rain. It's a lot of rain. Like maybe fifteen to twenty inches in a day's time, if you're lucky. Oh, and lots and lots of wind. As the hurricane descends upon you, that's when you get the **storm surge,** a ridiculous rise in water level by at least five or six feet in a Category 1 or 2 and twenty feet in a Category 5. (Let's hope, though, that you have the common sense to not ever be anywhere near a Category 5 hurricane. That's what God created cars and handguns for.)

The rise in water level from the surge, combined with the heavy winds and the waves they bring, is probably the most destructive part of the hurricane, causing severe flooding and crazy property damage. The water comes pouring over your feckless, meager levees like nothing, like it wasn't even there, like you wasted several days hauling sandbags around like an idiot for no good reason. Because now the water's coming over and you can't outrun it, and it takes you and it treats you like it caught you stealing a dirty magazine from its news rack. It just takes you and it hurls you around, smashing you into trees and cars and a shopping cart that was left in the street for some reason. If you can get up from that, you need to get to high ground fast. Any high ground you can find. That tree over there? Climb it! But be careful, hurricane winds have been known to send two-by-fours right through trees. Even if you manage to dodge the two-by-four, do you think the wind is gonna be any more gentle with you?

If you do keep your head and you're wearing your lucky T-shirt, you just might make it out alive. But do keep in mind, if the clouds suddenly vanish and the winds abate, the storm might not be over just yet. One morning in 1926 a bunch of people new to Miami Beach, Florida, experienced something very similar. The winds that were rocking their nouveau riche hurricane gala all night had stopped. Just stopped. Everything was calm and beautiful. Unfamiliar with hurricanes' tricks, many of them, ignoring the fevered warnings of a local meteorologist, wandered drunkenly down to the beach to admire the picturesque ring of dark clouds surrounding them. Some of them even waded into the warm ocean water for a morning swim. Within minutes the calmness of the hurricane's eye, in which they had momentarily found themselves, passed them by, and they were instead confronted with the violence of the eye wall. More than one hundred people never made it back to their party, or anywhere else for that matter. But you don't have to make the same mistake. Take this opportunity to jump down from the tree and get yourself into a sturdy house with several floors. But don't pull the surfboard from the wall it was rammed through and try to get all Dogtown on the waves right now. It's not worth it.

Hurricanes have caused a lot of damage to human society over the years. They've also caused us a decent amount of confusion. Like the live pig that one wedged into a tree fifteen feet up in a tree. Why and how? Or the swarm of thousands of butterflies another carried out into the middle of the South Pacific, where they flew over and past a boatload of confused sailors. Was that necessary? Or was it just Weather's idea of fun?

Maybe the weirdest thing that a hurricane has ever done was back in North Carolina in 1876, when the winds lifted *an entire church* up from its foundations and carried it floating down the road, around a corner, to another location in town, which was—get this!—the place where the parishioners had wanted the church built in the first place. It's popularly known as the Church Moved by the Hand of God. It's a very sweet story. Much nicer than when, twenty-four years later, that same hand of God killed six thousand citizens and razed the town of Galveston, Texas, as though it were Jericho in what has long been considered one of the worst disasters—natural or otherwise—to befall the United States. That's a less sweet story.

You Say Tomato, I Say Tropical Cyclone

Strictly speaking, a hurricane is not actually called a hurricane. Or it is, but not really. It's confusing; hurricanes go by a lot of names, and what each one is called depends on where it's striking. In the Atlantic and eastern Pacific oceans, yes, they are referred to as hurricanes, which comes from the Taino word *hurukán*. The

Tainos were the indigenous people of the Bahamas and other Caribbean islands, which put them smack-dab in the center of hurukán central. And given the fact that the word, to them, means "god of evil," it's not hard to guess what their feelings were on the matter.

Anyway, as long as you're reading this book somewhere around the Atlantic or eastern Pacific, I'm in the clear. I don't need to apologize to anyone or issue any corrections. But if you happen to be on a boat in the Pacific and you drift too close to Japan or the east coast of Asia, well then, you're just going to have to mentally replace the word *hurricane* with the word *typhoon* (from the Chinese word *daaih-fung* for "big wind") every time you see it printed. It's not my decision; it's global law. If it gets too confusing, you have my permission to use a pen to make line edits for yourself. But no doodling.

Oh, wait! You didn't start making changes with a pen yet, did you? Because you probably should have used a pencil. One with an eraser. You see, if your boat continues drifting and you end up in the Indian Ocean off the southern coast of Asia, you're going to have to erase all those *typhoons* and replace them with *cyclones*. And if you drift farther south to the Philippines, you have to change them to *baguios,* or even farther south to the coast of Australia, you'll need to make them into *tropical cyclones.* Come to think of it, that's what I should have been calling them to begin with. I mean, that is the official agreed-upon global standard name anyway. If you say tropical cyclone anywhere in the world, people will know exactly what you mean. From here on out, for the rest of the book, I'm going to call them tropical cyclones.

However, I just remembered I live in America. So whatever.

THE SAFFIR-SIMPSON SCALE FOR OH $%!#, IT'S A HURRICANE! RUN!

CATEGORY	PRESSURE	WIND SPEED	STORM SURGE	DAMAGE
1	28.94 in.Hg and up	74–95 mph	4 to 5 ft.	Defoliation of shrubs, trees, and ents.
2	28.50 to 28.93 in.Hg	96–110 mph	6 to 8 ft.	Small trees uprooted; mobile homes made less classy.
3	27.91 to 28.49 in.Hg	111–130 mph	9 to 12 ft.	Large trees uprooted; mobile homes made less existent.
4	27.17 to 27.90 in.Hg	131–155 mph	13 to 18 ft.	Windows blown; doors destroyed; roofs torn asunder; roads newly paved with water.
5	27.16 in.Hg and lower	156 mph and up	19 ft. and up	Strong buildings severely damaged; less strong buildings demolished; shards of mobile homes relatively unharmed; roads severely paved with deep water.

Santa Ana Winds

Before moving on to colder climes, I think we should stop for one brief moment to dispel a bit of conventional wisdom. When people consider the worst of what Weather has to offer, they usually think of conditions that are (a) wet, (b) cold, or (c) highly acidic.

If I asked you to list off some terrible awful things in the world, you'd likely come up with stuff like genocide, pandemics, animal extinctions, ska-funk fusion bands, serial killings, etc. But what are the odds that you'd include warm, dry winds? Seems unlikely, doesn't it? But little do you know. They can be just as devastating as the experience of watching some heavily tattooed nineteen-year-old mangle your favorite Toots & the Maytals song. Case in point: the **Santa Ana winds.**

Forming as an invisible lake of high pressure air above the hot, high, and dry plateau deserts of the American West's Great Basin, they spill downward, easterly and northeasterly, through the canyons of the San Gabriel and San Bernardino mountains of Southern California (among them is the Santa Ana Canyon from which they seem to get their name).

They gain momentum and heat and lose humidity, as they're compressed while on their way through the canyons. As they rush along, they carry sand and dust and steal moisture from the plant life as they roll, making for perfect brush fire conditions. In the autumn following a particularly dry summer, all you have to do is stop your car to burn one offending CD on the side of the road and you could be sparking a huge inferno. The dry conditions on the ground help the fire spread, and the strong winds themselves push it along toward residential areas.

In October 2007, 85 mph Santa Ana winds carried a series of wildfires across Southern California, into Los Angeles, San Diego, and San Bernardino. Five hundred thousand acres of land were scorched, and more than fifteen hundred homes were destroyed. Even with help from the army and the National Guard, it took several weeks for all the fires to be extinguished. The only plus side to the whole affair was that a number of ska-funk fusion albums under production at the time were abandoned amid the chaos.

Ice and Snow

Have you ever noticed that sometimes it's simply too cold to snow?

No, you haven't. Because it's not true. It's never too cold to snow. That's a myth propagated by radical anti-frozen-precipitationists. To what fiendish end, we can only speculate. And speculate we will. I think they're probably in league with Weather itself—and maybe the libertarians—to keep us all off guard and fearful. Break your mind free of their insidious propaganda, and consider the facts.

While it is true that cold air is less capable of holding moisture and therefore usually more dry than warm air, the relative humidity never drops so low under natural conditions to keep precipitation from occurring. Especially since it's easier for cold air to reach a relative humidity favorable to the accumulation of cloud droplets. (As you should remember from Chapter 4, snow and rain form the same way, with raindrops often beginning their lives as snowflakes before melting on their way to the ground.) So even though the dry air associated with very cold temperatures may keep snow from falling, it's due to the dryness, not the coldness. Still, I guess it's true that it snows less often under very cold conditions, but it's certainly not true that it *can't* snow under very cold conditions.

You know what else the radicals would have you believe? That snow shouldn't occur when surface temperatures are above the freezing point. Another untruth! But this one's a little more complicated to dispel. To do so, we'll have to draw upon our previously acquired knowledge that in order for water to change states, it needs to expel or extract energy. So what would happen if a snowflake fell from a cloud and into 36°F air, a little warmer than its melting point? It should melt, right? Actually, here, yes, you're right. The snowflakes begin to melt, but in doing so, they have to suck in heat energy from the surrounding air to get their water molecules sufficiently jumpy and agitated enough to make the leap from the frozen to liquid state. Now, with less heat energy in it, the surrounding air begins to cool down, which ironically helps to keep the snowflake from melting before it hits the ground. Even though it's highly unlikely, it's actually not impossible to find snow falling in temperatures up to 50°F. But that's not what the anti–frozen-precipitationists would have you believe, now is it?

Oh, their spin machine is good. I'll bet you've also been duped into believing that snow is always white, haven't you? Well, for your information, there have been numerous incidents in which snowflakes of color have been reported falling. Case in point: the pink snowstorm that fell over Colorado in 1895. And the blood-red snow that turned Paris momentarily crimson in 1810. Then there's that blanket of yellow snow that covered parts of Central Europe in 1991 and the brown snow that drifted down upon Victoria, Australia, in 1935. And how could we forget the black snow that fell over New York in 1889, Arkansas in 1940, and a bunch of other places in between? Yes, that's right, black snow! You probably thought that was impossible, didn't you? Well, it's more than possible; it happened. Black snow power!

Different-colored snows each have a different reason for coming into being, and it's due to outside elements mixing with the snowflakes (but it's important to remember that that doesn't make them less legitimate somehow). For example, pink or red snow could be caused by the red dust of nearby dust storms getting kicked up into the sky and comingling with snowflakes as they form in clouds. Or as Charles Darwin himself discovered when he came upon a red snowfall in the Andes, it might be due to the presence of tiny bits of red-tinged algae that have become airborne. Yellow snow, similarly, might be attributed to microscopic pieces of pollen or clay in the air, while brown snow could be fine particles of sand. And black snow is very likely the result of ash dispensed from a fire in the vicinity. Open your mind, man.

Oh, and there's just one other thing that we need to clear up before going any further. You know conventional "wisdom" about there being no such thing as two identical snowflakes? You know what I'm talking about. Anyway, that is correct actually. Or I should say, it's correct at least as far as we know. Obviously, we can't observe every snowflake that falls from the sky under a microscope, but people have observed a whole lot of them. Thousands and thousands. Wilson A. Bentley, a Vermontinarian (or Vermontineer, or whatever; he lived in Vermont), went his whole life photographing snowflakes every chance he could. And not one of them matched any other of them. Once some snow scientists— or *snientists*—in Colorado thought they found two that were exactly the same. But upon further research it turned out they weren't exactly the same enough, which makes them not exactly the same.

Ice Storms

Do you know what does legitimately suck? And I think this is something over which we can put aside our grievances with the anti–frozen-precipitation people. Ice storms.

They suck. Oh, man, do they suck! How much do ice storms suck? A whole freaking lot. That's how much ice storms suck.

If you don't believe me, then you don't live in a place that is prone to ice storms. Because if you did, you'd be right there with me on this one. What happens is you wake up in the morning and you look out your window, and everything—everything, every possible surface—is covered in a layer of glistening, gleaming, twinkling ice. Does that sound beautiful to you? Well, you're wrong. It's not beautiful. It's disgusting. All the trees and bushes in your yard sag under the weight of the ice. Power cords and telephone lines, also sagging to their breaking point, often reach that breaking point and snap, so electricity and phones go out. Walking out your front door and down your porch steps is suddenly the most difficult, dangerous, impossible thing you've ever risked your life doing ever. And if you do make it to your car without slipping on the slickness of your driveway and busting your coccyx, you might as well forget about it and turn around and risk your life trying to get back inside your house, because you don't want to drive anywhere. You don't want to put that car on any road that has any other cars anywhere near it. If you're driving along on an ice-covered road and you decide that you need to make a turn, just save yourself the effort of turning the steering wheel and ask your car politely to turn for you. You'll get roughly the same results.

So, from where does this horrible agent of Weather originate? Again, it's essentially the same thing as rain, with a twist. If surface temperatures are below freezing but the temperature in the storm cloud isn't cold enough to produce snow or if there's a layer of warmer air between the cold cloud and the cold surface, any precipitation that falls to the surface will freeze before it hits the ground. It'll freeze into a tiny little ice cube, usually less than 0.25 inch in diameter, that clicks against the street and your car hood and your bedroom window and then bounces away to accumulate on the ground. This is **sleet,** and if you encounter sleet, you're off the hook. Sleet's not that bad, all things considered.

However, if the colder surface air is too shallow for the raindrop to freeze up before it hits the ground, it'll splatter when it comes into contact with anything and immediately freeze into a thin veneer of ice. This is **freezing rain,** and this is the stuff that sucks. Anyway, that keeps happening over and over again until the thin veneer of ice becomes a thick veneer of ice, and before you know it, you're watching helplessly through your windshield as your car slides into a mailbox designed to look like a barn for teddy bears. And then, even if you try to drive away real quick, you can't because your tires aren't getting any traction, so when the teddy bear lady comes outside to see who destroyed her favorite wood-working project, you have to pretend you were just getting back into your car to get your insurance papers from the glove compartment. She'll act like she believes you, but she doesn't really, and she'll keep looking at you suspiciously throughout the information-exchanging process. Three hundred dollars for a teddy bear barn? For that thing? What a steaming load of crap!

So yeah, ice storms suck.

Hail

What of **hail,** the huge chunks of ice that fall from the sky? Surely they must be the result of some extremely cold temperatures. That would be absolutely correct, if you didn't understand the meaning of the word *correct*. Against all reasonable logic, hail actually forms as a result of warm temperatures. In fact the warmer the temperature, the larger the chunks of ice will grow. Doesn't make sense, does it? Well, once you better understand the way a hailstone forms, it'll make more sense.

Hail is a phenomenon directly related to both the height of the cloud in which it forms and the force of the updraft within it. They tend to be closely associated with severe thunderstorms. I'm going to ask you again to try to recall from Chapter 4 how raindrops and snowflakes form. Inside a tall cumulonimbus cloud there's a huge temperature difference between its base, relatively close to the ground, where the air is warmer, and its top, pressing up against or pushing into the much colder stratosphere. Up at the top of that cloud, where the cloud droplets are supercooled, raindrops can freeze into little bits of ice called graupel. (Hey, remember earlier in this chapter when we discovered that hail and graupel help cause lightning? See, we're doing a whole cyclical thing here. What, that doesn't impress you? Ah, whatever . . .) When that graupel is held at that altitude or propelled back up there by strong updrafts, more and more supercooled cloud droplets freeze onto them, and they get bigger and eventually form small pieces of hail.

If the updraft isn't all that strong, the small hailstones can use their slight accumulation of weight to drop down and out of the cloud. In that situation, they'll either reach the surface about the size of sleet and without much fanfare. No shattered windshields or broken roofs or screaming, running people. Or even less excitingly, they will melt on their way down and fall to the ground as run-of-the-mill annoying rain.

However, if the updrafts are sufficiently strong, the hailstones remain suspended for several minutes at a time. Updrafts like that require very unstable air with surface temperatures much warmer than the air aloft. Remember, warm air always wants to go up, but if there's warm air above it, that air also wants to go up, so there's not really anywhere for the surface air to ascend. But if the air above it is cooler, then it wants to come down, and the warm air can rush right up as a strong updraft. And in that situation a hailstone much heavier than you would probably guess can remain suspended. Or it might fall down to the lower part of the cloud only to find itself propelled back up by the strong updrafts.

While all this is happening, as the stone passes through cloud droplets, it continues to collect moisture on its surface. Up top, supercooled particles are freezing directly onto it. Down lower, it's getting coated in liquid, which will freeze to it once it travels back up to the top. And every time this happens, it acquires a new layer and gets bigger and heavier. If it remains suspended for somewhere between five and ten minutes, it will have collected about ten billion cloud droplets and will have grown to the size of a Titleist Pro V1x golf ball, which would impress any golf fan—if it doesn't fall and land directly on his skull. Because by this point these things are dangerous.

Can you imagine a golf ball–sized piece of ice hurling downward at your head at around 100 mph? That's roughly the speed at which it will fall if it should break free from the updrafts now. Just think of the damage it would do. And now think of a whole bunch of golf ball–sized pieces of ice all hurling loudly toward the Earth at the same time, the sound caused in part by electrical charges being exchanged between them as they descend. And now think about how that would make you feel if you happened to be, say, out at the thirteenth hole with no shelter anywhere close by.

Hailstorms cause a billion dollars in damages to property and crops in the United States every year. They pummel plants, destroy cars, ruin roadways, shatter windows, bore through roofs. In 1990 one hailstorm alone, boasting stones the size of baseballs, caused more than six hundred million dollars' worth of destruction to the Rocky Mountain area of Colorado.

They also kill things. Living things. Like wild animals and farm animals and pet animals and people animals. In Nebraska in 1917 one terrible storm killed hundreds of chickens and mutilated some number of horses and cows that had been left outside. They were later found broken and bloody, covered in welts. In 1978 a flock of two hundred Montana sheep met a grim fate when one such storm occurred. They've also been known to knock birds right out of the sky. A storm of plummeting ice left the Louisiana State University campus covered in avian corpses. In 1979 a baby in Colorado was struck and killed, and in 2000 a grapefruit–sized hailstone connected soundly with the head of a nineteen-year-old man who was attempting to move his car from harm's way. The United States has gotten off relatively easily. In 1996 ninety-two people were killed utterly to death during a massive storm in Bangladesh.

The U.S. high plains may think they've got it bad, as that's where the country's storms happen most often, featuring the largest stones (the size of volleyballs once in Nebraska in 2003), but they don't have anything on Bangladesh or northern India. People get killed by hailstones there *all the time*. As the result of one hailstorm in the Indian Moradabad and Beheri districts back in 1888, a shocking 246 people were killed. One out of every four hailstones that fall there measures at least an inch in diameter, whereas only one in twenty reaches that size on our high plains. (Why does God hate that part of the world so much?)

We can't move on from this section on a note like that. So, uh . . .

Knock knock.

(Now you say, "Who's there?")

Hailstone that reportedly fell in Germany in 1896.

(You say, "Hailstone that reportedly fell in Germany in 1896, who?")

Hailstone that reportedly fell in Germany in 1896 that was found to contain an entire frozen fish.

I love that one. But I'll bet you thought I was going to say, "Hailstone that reportedly fell in Germany in 1896 that was found to contain a living frog." Didn't you? No. That happened in Iowa in 1882.

CHAPTER 6
READING THE SKY

So, if you've read this far, you are, for all intents and purposes, a preeminent expert in the scientific study of meteorology. And if you've skipped past the boring parts, you're a junior preeminent expert. That's good enough. Congratulations. Pat yourself on the back. But don't take all day doing it, because there's still lots to cover. And this is the part of the book in which we finally engage Weather on its own turf.

We've followed Weather from the surface of the Earth up to the highest reaches of the atmosphere. We've tracked it as it made its roundabout course from the equator to the poles and on its roller coaster ride through its jet streams. We've observed its behaviors through a wide-angle lens and a microscope. We've marched with its armies along the wind-scarred battlefields of the middle latitudes and climbed up inside its beast of war. There is naught to do now but engage it head-on.

Weather fights to control us through a tyranny of inclemency, and we must beat it back and retain our God-given freedom. We as a species have yet to possess the power to destroy Weather utterly. But if we can presage its maneuvers, keeping our gait one step ahead, we may be able to hold it at bay until some brave scientist designs an effective surface-to-Weather missile.

With that in mind, our first mission is to break Weather's code. And where is that code written? In the sky, of course. And what is the most prominent feature of the sky? The clouds.

Tiers of a Cloud

The Four Families

Cirrus: These clouds sit high in the sky, anywhere from four to seven miles up. Because of the frigid temperatures up there, they're made up primarily of ice crystals. Don't bother trying to grab hold of one, even with a ladder; they're way too high up there.

Alto: Middle-layer clouds. These can be found somewhere in the one- to four-mile range in the atmosphere and are made up of both water droplets and ice crystals. Their voices are ideal for compositions written in the range between G below middle C and the E a tenth above middle C.

Cumulus: You know those puffy clouds that angels sit on to play their harps? These are them. Definitely the cutest clouds of the bunch. Or the most ominous. Depending on what song the angels are playing.

Stratus: Overcast clouds. These are the ones that make it impossible to see the stars while you're on vacation in the mountains, even though you only get to go up there once a year, and—goddammit!—you need to see the stars, like, once a year at least. The city lights make that impossible the other fifty-one weeks out of the year. Can't you get just one lousy week of stars? Is that really so much to ask? Anyway, that's caused by stratus clouds.

High Clouds

These are the **cirrus** clouds mentioned above. Since they're more than four miles up, their thin, sheetlike shapes are greatly influenced by upper-altitude winds that pull them along as they blow. Although they might portend rain or snow, they don't cause it themselves

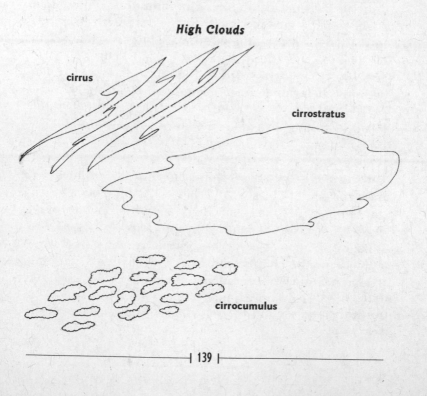

High Clouds

cirrus

cirrostratus

cirrocumulus

and are rarely thick enough to block the sunlight to any great degree, so shadows will still be cast.

Cirrus (Ci) **Cirrus** clouds, the patriarch of the cirrus family, are the most common and generic, and hence the entire family draws its name from them. They're thin, little wisps of loose cotton, floating about on their own, without much heft or billowing qualities, as though a gigantic mattress had been torn apart in the sky. During the day they're usually white, but at sunrise and sundown, they may appear red, orange, or yellow as sunlight pours through them from a lower angle. These are pretty much the most boring clouds.

Cirrocumulus (Cc) Like small, roundish dollops of mayonnaise high up in the sky, **cirrocumulus** clouds float in the upper atmosphere, either by themselves or in long rows, making everything delicious. If you catch a bunch of them together, they may have a rippled effect, like the scales of a fish, which led to the term *mackerel sky* describing their appearance. *White fish sky* would probably describe it more accurately, but presumably the term was coined by some guy who was really into mackerel, and now the case is closed.

Cirrostratus (Cs) These clouds form a very thin sheet over the top of the sky, like ladies' diaphanous undergarments of the finest silk. As the sun shines through these unmentionables, its light gets refracted, forming a halo around its edges. If not for this halo, you might not even know the clouds were there, delicate as they are. The appearance of **cirrostratus** clouds in the sky may signal a storm approaching sometime within the next twelve to twenty-four hours. This is especially true if they are followed by either of the two cloud types in the next category.

Middle Clouds

Both these two cloud-types belong to the **alto** family, with temperatures ranging from 32° to –13° F. The water droplets within these clouds are very small, and even though they exist below the freezing point, they won't necessarily freeze into crystals. Clouds at this altitude often contain too little dust, ash, or pollutants for water droplets to latch on to and form ice.

However, when an airplane flies through them, its exhaust may provide the tiny bits of stuff for crystals to form. That's why you'll often see long strings of white clouds trailing behind a jet as it flies through the sky. These are called **contrails.**

Altostratus (As) These are gray or blue-gray clouds comprised of a mixture of ice crystals and water droplets. Like cirrostratus clouds, **altostratus** clouds can

Middle Clouds

altostratus

altocumulus

cover the entire sky for hundreds of miles around. Through these dense, thick clouds, the sun and moon lose a lot of visibility and can be made out only as round, bright, blotchy areas in the sky. If you're looking up at a murky sun set within a depressingly gray sky that kind of reminds you of a Dostoevsky novel, you're probably looking up at an altostratus cloud. If you can make out a halo around the sun, it's probably a cirrostratus. If you're still confused, take a look at the ground. If you see your shadow, it's more likely a cirrostratus. (This does not work at night.)

Altostratus clouds can often be found in the sky as a warm front approaches, so you can make a pretty good bet on long and lingering but generally mild rain or snow arriving within the next couple of days. So don't plan any picnics in the short term.

Altocumulus (Ac) These are made up mostly of water droplets, and they tend to be rounded, fluffy, and gray with some areas darker than others. Often you'll see a bunch of them together in the sky, not quite blocking it out but keeping a steady stream of sunlight from hitting the ground. Each individual **altocumulus** cloud is a bigger, dirtier dollop of mayonnaise than the cirrocumulus clouds.

If these come hanging around your neighborhood on a warm and humid summer day, go buy yourself an umbrella because you've probably got some thunderstorms a-comin' your way.

Low Clouds

The clouds that hang about a mile or two up in the air are usually made up entirely of water droplets. However,

Low Clouds

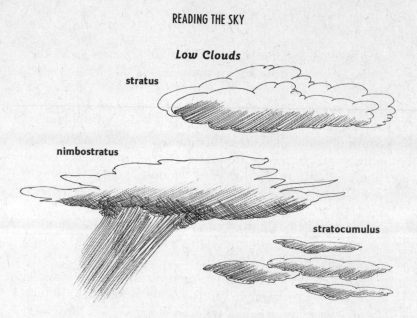

stratus

nimbostratus

stratocumulus

if temperatures are cold enough, they may also contain ice crystals or even snow.

Stratus (St) Like the cirrostratus and altostratus, **stratus** clouds (from which their family takes its name) cover the entire sky, prompting everyone beneath them to question the meaning of life and the existence of absolute morality. It's like a fog that doesn't quite ever touch the ground, and in fact in many cases they are caused by fog that has lifted. These block out the sun almost completely but will give forth neither rain nor snow. They might, however, decide to spray you with a cold mist or drizzle that screws up your eyeglasses and hinders your vision no matter how many times you wipe the lenses on your shirt.

If rain does begin to fall, then what you've probably got yourself are some **nimbostratus** clouds. You can also tell the difference by looking at the bottom of the cloud. If you can make out its base, it's a stratus cloud.

Nimbostratus (Ns) These are the ultimate rainy day clouds. They cover the sky in a dark gray wet blanket of continuously falling rain or snow, completely blocking out the sun or moon. It's not even the kind of precipitation that patters against your windows while you sleep. Not even that interesting. Instead it just falls gently and steadily. It's there when you go to bed, and it's still there when you wake up. The kind of rain that belongs in a very thoughtful and atmospheric art film. Definitely the worst kind of rain.

Sometimes, if there's enough moisture in the air, a layer of clouds even lower than the nimbostratus will form. These are **stratus fractus** clouds. They form and dissipate quickly, with winds giving them a ragged, torn look. Given the appropriate amount of ground heat and unstable enough air conditions—wherein cold air, wanting to move downward, rests on top of warmer air, wanting to move upward—they may develop into **cumulus** clouds.

Stratocumulus (Sc) These are low-floating, puffy clouds that may be found in rows (like cirrocumulus clouds) of either light gray, dark gray, or lightish dark gray, but rarely give up much rain. Blue sky can often be seen peeking out from behind these clouds to check in and make sure everything's still all right down there. When God needs to make a big show of things, he uses stratocumulus clouds through which to shine shafts of glorious sunlight. It's a really cool effect—kind of his calling card.

If you're trying to distinguish them from altocumulus clouds, you may notice that they hang much lower in the sky and each puff is bigger—about the width of your fist. (Not actually; just from your perspective. Don't be stupid.)

Vertical Clouds

These clouds can exist in all three levels of the sky at once, forming from the bottom up as hot air radiates upward from the ground, sometimes as fast as 100 mph, in unstable air conditions. In just a few minutes one of these babies can grow from a cute white fluffy puff into a dark and terrifying thunderhead. Don't make these clouds angry. You wouldn't like them when they're angry.

Vertical Clouds

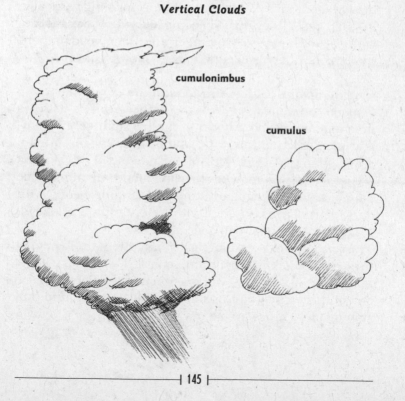

cumulonimbus

cumulus

Cumulus (Cu) The white or light gray **cumulus** clouds (from which the family takes its name) may take any of a countless number of shapes. They can look like a fluffy cotton ball, a fluffy football, a fluffy race car, a fluffy ottoman, a fluffy giraffe, a fluffy six-pack, a fluffy lower Florida, a fluffy Abraham Lincoln, or even a fluffy Starship *Enterprise* (NCC-1701-C). They usually have sharp, fluffy features and a flat base. On a warm, humid day, they may float no more than a half mile above the ground (which is still too low for you to reach up and grab, but is pretty low all the same). They are generally associated with calm, fair conditions. Nothing to fear here.

Unless . . . warm temperatures beneath and unstable air conditions cause them to start forming vertically. In that case, they may become **cumulus congestus,** with their shapes growing taller and their heads mushrooming out. If this occurs, expect showery rain.

Cumulonimbus (Cb) A **cumulonimbus** cloud may hang a mere two thousand feet above the ground, with its height going all the way up to the tropopause. It's usually long, tall, and thin with its top flattening out into an anvil shape as high-altitude winds send it blowing across the sky. It may be traveling alone. Or it might be part of a wall of similarly dreadful thunderheads with dark, warm bases bursting with water droplets and cold, windy tops full of swirling ice crystals.

You can expect practically anything from one of these cumulonimbusis: rain, snow, hailstones, lightning, thunder, frogs. If you open your front door and see one of these guys waiting for you, just turn around and stay inside. Things are gonna get ugly.

Cumulocongestus clouds may be mistaken for cumulonimbus clouds at times, but you can usually pick them apart by observing their tops. If you can make out the edges of the top, it's probably *not* a cumulonimbus. Another clue: If you hear rumbling thunder or a bolt of lightning strikes a tree in your front yard, setting it aflame, it *is* a cumulonimbus.

Those are the ten main types of clouds, and if you look up in the sky and determine what you're looking at within that framework, you're well off on your way to beating Weather at its own game. There are, however, a number of subcategories of clouds with interesting enough characteristics.

For example, **lenticular** clouds are forms of cirro-, alto-, or just plain cumulus clouds with elongated widths and sharply defined outlines. Stupid people will sometimes confuse them with spaceships. (To be fair, though, stupid people are apt to confuse most things with spaceships.)

Mammatus clouds occur as subcategories for many types of clouds, including cumulonimbis and altocumulus, and are recognized by how much they seem to hang just perfectly from the sky. Their name comes from the word *mammary*. People never get tired of looking at these clouds for some reason.

Castellanus clouds take their name from the word *castle* and do indeed resemble castles in the sky. They're long and tall with towerlike extensions, often accompanied by **draculanus** and **vanhelsinganus** clouds. These two smaller cloud types, often seen chasing each other across the sky, resemble a young Christopher Lee and Peter Cushing respectively and do not in fact exist. Although they should.

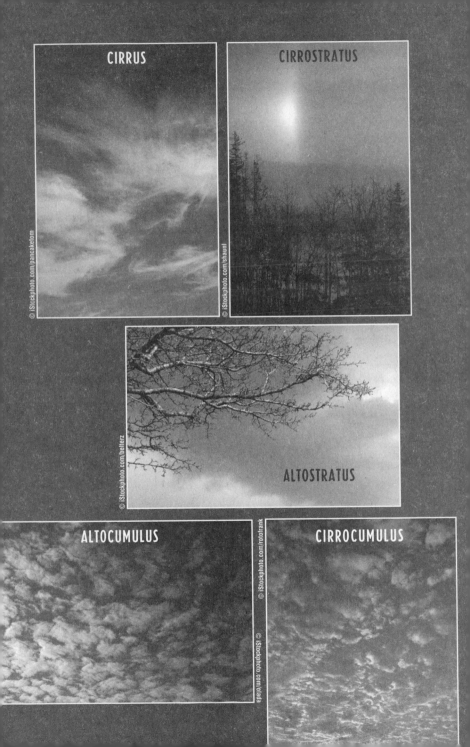

STRATUS

© iStockphoto.com/A330Pilot

NIMBOSTRATUS

© iStockphoto.com/guysargent

STRATOCUMULUS

© iStockphoto.com/azgek

CUMULUS

© iStockphoto.com/rotofrank

CUMULONIMBUS

© iStockphoto.com/DHuss

Predicting the Imprecise Improbably

Now that we're able to discern which kind of cloud is which, and we understand the behavior of different air masses and fronts, we can move on. Put the pieces together and see if we can't try and guess at what Weather has up its sleeve, on the basis of what we see.

For example, let's say you were watching the sky diligently, and you noticed the wispy little cirrus clouds floating about on their own way up high in the blue sky for two days straight. Do you have any idea what that would mean?

No, you don't. Even if you think you do, you don't. Cirrus clouds can come in advance of a cold front or a warm front. I'm sorry; that was a trick question.

However, if you did happen to be paying attention to the wind, it might just give you a hint. If the winds are consistently blowing from the *southwest, west,* or *northwest,* you may have some brief but turbulent storms heading your way within the next day or two, followed by a sharp drop in temperature. We can guess this because we know those winds to be associated with cold fronts.

And if the winds were steadily moving in from the *south* or *southeast,* we might guess that for the next four or five days you'll have pretty mild weather, with the skies slowly growing darker and your sinuses getting progressively more agitated until a not particularly violent rain sets in, reaches for the remote control, and makes itself comfortable a few more days afterward. Once that ends, we could reasonably expect warmer weather. Again, we know that those winds are associated with warm fronts.

While we're on the subject, it turns out there's actually a lot that we can actually glean from the winds.

Unpleasant conditions generally tend to travel in on *northeasterly* winds, while much nicer days come in on *northwesterly* winds. If you wake up and it's raining, and the winds are *northeasterly* or any direction down to *southerly* but then shift suddenly and begin to blow *westerly,* those rain clouds are getting blown away, and the sun is on its way back. Conversely, if you're looking up into the sky and it's cloudy, and the winds then shift from *southwesterly* to *southeasterly* or from *northeasterly* to *northwesterly,* expect sudden and vicious winds and rain, followed by more winds and rain; a squall is likely heading your way, followed by a big ol' cold front. And if the wind turns suddenly from any direction and begins blowing straight down on you, expect a gigantic country-sized meteor to land on you within minutes.

But anyway, those wispy cirrus clouds. Let's say that you get bored of looking at them and go watch television for a while. Then, when you go outside again to get the mail, you notice that they've been replaced by fish scale–looking cirrocumulus clouds. That means they're bulking up; much more coverage than before. Then you go away, and when you come back again a few hours later, those fishy clouds are gone, and there's the much lower and darker altocumulus clouds in their place. Man, those clouds are descending fast. Something's happening. Can you guess what it is? (Here's a hint: It's gonna get cold.)

If this is in fact a cold front coming, most likely these clouds will quickly be replaced by the much lower, dark, and blotchy stratocumulus. If that happens, it's going to rain. The clouds are getting closer to the ground as the pressure drops and the cold front gets nearer. Af-

INSTANT WEATHER FORECAST CHART

SEA LEVEL PRESSURE	PRESSURE TENDENCY	SURFACE WIND DIRECTION	SKY CONDITION	24-HOUR WEATHER FORECAST
1023 mb or higher (30.21 in.)	Rising, steady or falling	Any direction	Clear, high clouds, Cu	1, 18 (in winter 14)
1022 to 1016 mb (30.20 to 30.00 in.)	Rising or steady	SW, W, NW, N	Clear, high clouds or Cu	1, 18
	Falling or steady	SW, S, SE	Clear, high clouds	1, 3, 17, 5
	Falling	SW, S, SE	Middle or low clouds	6, 17
	Falling	E, NE	Middle or low clouds	6, 14
	Falling or steady	E, NE	Clear or high clouds	3, 5, 14
1015 to 1009 mb (29.99 to 29.80 in.)	Rising	SW, W, NW, N	Clear	1, 14
	Rising	SW, W, NW, N	Overcast	2, 16
	Rising	SW, W, NW, N	Precipitation	11, 2, 16
	Falling	Any direction	Clear	3, 17 (dry climate, summer 1, 15)
	Falling or steady	SW, S, SE	High clouds	3, 17, 5
	Falling	SW, S, SE	Middle or low clouds	7
	Falling	E, NE	Middle or low clouds	7, 12, 14
	Falling	SE, E, NE	Overcast, precipitation	9
	Falling	S, SW	Overcast, precipitation	10, 13

SEA LEVEL PRESSURE	PRESSURE TENDENCY	SURFACE WIND DIRECTION	SKY CONDITION	24-HOUR WEATHER FORECAST
	Rising	SW, W, NW, N	Clear	1, 12
	Rising	SW, W, NW, N	Overcast	2, 12, 16
	Rising	SW, W, NW, N	Overcast with precipitation	11, 12, 16
	Rising	NE	Overcast	4, 12, 13, 14
	Rising	NE	Overcast with precipitation	11, 12, 13, 14
1008 mb or below (29.79 in.)	Rising or steady	SW, S, SE	Clear	3, 6, 8, 12, 15
	Falling	SW, S, SE	Overcast	7, 8, 12, 13
	Falling	SW, S, SE	Overcast with precipitation	8, 10, 12, 13, 16
	Falling	N	Overcast	4, 14
	Falling or steady	E, NE	Overcast	7, 12, 14
	Falling	E, NE	Overcast with precipitation	8, 9, 12, 13

WEATHER FORECAST CODE

1 . . . Clear or scattered clouds

2 . . . Clearing

3 . . . Increasing clouds

4 . . . Continued overcast

5 . . . Precipitation possible within 24 hours

6 . . . Precipitation possible within 12 hours

7 . . . Precipitation possible within 8 hours

8 . . . Possible period of heavy precipitation

9 . . . Precipitation continuing

10 . . . Precipitation ending within 12 hours

11 . . . Precipitation ending within 6 hours

12 . . . Windy

13 . . . Possible wind shift to W, NW or N

14 . . . Continued cool or cold

15 . . . Continued mild or warm

16 . . . Turning colder

17 . . . Slowly rising temperatures

18 . . . Little temperature change

Originally published in Meteorology Today

ter that, you can expect the clouds to spread out and cover the sky in darkness as the bummer-inducing nimbostratus or to build up in height as the fearsome bad-ass cumulonimbus. And that's when the rain comes. But as we know, it probably won't last long. But it will leave much colder temperatures in its wake. Which could be nice if it happens to be a disgustingly hot summer day. But if it's winter . . . go dig your scarf out of the closet.

What say we back up, rewind the tape, and return to the wispy cirrus clouds again? This time they're replaced by the fishy cirrocumulus clouds again, but these clouds last a lot longer. Like the better part of a day. The clouds descend and spread out into altostratus clouds. Your shadow disappears, and the sun is just barely visible through the thick cloud cover. This looks like an impending warm front.

Get your poetry-writing journal, because the next few days are most likely going to be ripe for angst. You can count on those clouds to descend slowly, lazily, like the giant palm of God descending upon you, into stratus clouds and maybe some stratocumulus clouds. Then come the deep, dark, all-encompassing nimbostratus clouds with their slow, merciless rain. The temperature will gradually rise, and the winds will turn. And the warm front will go slouching toward Canada, leaving you in its humidity.

But don't worry too much. A cold front is almost always just a few days behind the passing of a warm front.

Really, that's all you need to make relatively accurate predictions a day or two in advance. Just pay attention. Watch how Weather moves, how it turns and descends. Pay attention to the sequence of events. What follows what and all that. After a few weeks or months, you'll start to get the hang of it and you'll be able to look out your front door and then reach for your raincoat.

Something to keep in mind, though, is that no two fronts ever move through in quite the same manner. These are all rules of thumb, not rules of iron fist. Weather is far too predictably unpredictable to get cornered so easily.

To really poke your finger beneath the rib cage of Weather and twist it around, you're going to need to arm yourself with some weaponry, the same weapons that humans have been using for hundreds of years. They're easy to use once you learn the barrel from the butt. And not very expensive. If you like projects, you can even make your own.

Let's arm you with some instruments.

CHAPTER 7
THE RIGHT TO BEAR INSTRUMENTS

These are your instruments, and to you, they are everything. They are all that stands between you and Weather. You will love and respect and care for your instruments. You will give your instruments names, like Barry the Barometer and Suzy the Psychrometer, or better ones maybe; I'm bad with names. You will sleep with your instruments, but not actually sleep with them, like not in the same bed or anything, because then they wouldn't really be effective as instruments, and some of them are pointy. Okay, you will sleep in the same town as your instruments, which isn't really saying much, but they will know your true heart. Anyway, let's just get to the instruments you will need to defend yourself effectively against Weather, because this analogy isn't working as well on paper as it did in my head. I'm sorry.

The Thermometer

This admittedly obscure instrument can prove incredible useful, provided you're able to crack the baffling code of information it spits out. What a **thermometer** measures is the temperature of the air surrounding it, and it's usually in the form of a glass tube containing a thin portion of colored liquid. Running up the length of the glass tube is a series of glyphs that scientists call **numbers,** or units of mathematical measurement. The number at which the colored liquid comes to rest represents the air's temperature. When the colored liquid rests next to a low number, it probably means that the air surrounding the thermometer is cold. When it rests next to a high number, it means that the air surrounding the thermometer is hot. And when the thermometer melts, it means that the air around the puddle of molten glass is extremely hot, and it's probably time for you to burst into flames.

Here's how you use a thermometer: Look at it. That seems simple enough, doesn't it? But you'd be surprised by how many amateur meteorologists make the beginner's mistakes of trying to ingest it (through the mouth or variable other orifices), slap at it, or speak directly with the liquid inside. Such efforts will prove ineffective. All you need to do is look at it and read its datum.

Despite its extreme complexity, the thermometer actually works with a quite elegant simplicity. At the bottom of the tube is a small bulb that holds a good deal of the liquid. The air surrounding the bulb warms the liquid (usually alcohol or mercury), and as the liquid warms, it expands. As it expands, it is pushed upward from the bulb and into the tube. The warmer it gets, the more it expands and the higher up in the tube it rises.

Thermometer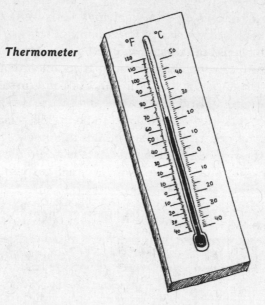

There's no use trying to reason with the thermometer. It's simply reacting to the temperature around it, not vice versa. Even if you crack it open and force the liquid upward or downward with your finger, it will have no bearing on the temperature around it.

Not all thermometers rely on the manipulation of liquids to determine the temperature. The **bimetallic thermometer** uses two strips of different types of metal (typically brass and iron) welded together. The brass expands faster in heat, so the strip bends more and more as the temperature rises. This does something to something else inside the thermometer, and eventually the temperature is displayed on a round clocklike face. These look like large sticking pins with clocks where the face should be and are the kinds of thermometer that people stick into turkeys on Thanksgiving to ensure that all the white meat is way too dry for consumption.

The **electrical thermometer** uses electronics to ascertain the correct temperature. Some electrical things happen, then a silicone chip kind of thing inside registers something, and then the temperature gets spit out on an LCD display. These tend to be highly accurate, on account of their utilizing electronics and such. (The least accurate type of thermometer is **Fred.** Fred will stick his arm out the window for you and then tell you whether

A VERY USEFUL TEMPERATURE CHART

TEMPERATURE IN FAHRENHEIT	TEMPERATURE IN CELSIUS	CONDITIONS
−40°	*−40°*	Extreme cold; everything frozen; pray for one of your limbs to catch fire.
0°	*−18°*	Too cold for you; go back inside; what are you, an idiot?
32°	*0°*	Water freezes; whiny bitches bitch.
64°	*18°*	Pleasant spring weather; optimum conditions for crafting insipid poetry.
85°	*29°*	Pleasant summer weather; go to the beach and try not to drown.
97°	*36°*	Disgusting summer weather; punch guy next to you for breathing.
106°	*41°*	Too hot, too hot; stay inside and plot suicide.
160°	*71°*	Forget it, you're already dead.
212°	*100°*	Water boils away on lifeless planet.

or not you need to wear a coat. His data are correct almost 50 percent of the time.)

The actual temperature may be read on one of two main scales: **Fahrenheit** or **Celsius.** On the eminently logical Fahrenheit scale, which is widely used in the United States, Belize, and some guy's basement in Turkmenistan, water freezes at 32° and boils at 212°, as any thinking person would assume. (Or is it 221°? No, it's definitely 212°. I think.) However, the Celsius scale has some other ridiculous ideas: It has water freezing at 0° and boiling at—get this!—100°. I'm sure it will come as no shock to learn that the anti-American rest of the world uses this confounding temperature scale, purely out of spite.

The thermometer is an invaluable tool for any budding backyard meteorologist.

The Hygrometer (or, If You Will, the Psychrometer)

A close cousin to the thermometer is the incredibly more popular and easily understandable **hygrometer,** which measures the humidity of the air around it. Probably the most user-friendly version is the **sling psychrometer,** consisting of two identical thermometers mounted next to each other on a piece of wood, metal, or bacon.

The bulb of one of the thermometers is wrapped in a piece of moist cotton or linen; this is the **wet-bulb thermometer.** The bulb of the other is kept dry in the open air; this is the **not-wet-bulb thermometer.** These two mounted thermometers are connected to a rotating handle or chain. To use it, you dampen the cotton around the wet bulb, pick the psychrometer up by the handle, and swing it around over your head like an idiot for a few minutes. (If any rival meteorologists approach you looking to steal your measurements, this will put you in optimal defense mode.)

The way it works is that the water around the wet bulb will evaporate, cooling the bulb beneath it. This causes the temperature on the wet-bulb thermometer to drop. The not-wet-bulb thermometer registers the actual temperature. By calculating the difference between the two readings, you should be able to calculate the **relative humidity** of the air. If the relative humidity is low, the air around the wet bulb will allow for a lot more evaporation, and the difference between the two readings will be significant. If, however, the relative humidity is high, less air around the wet bulb will evaporate, causing less of a temperature drop, and the difference in readings will be slight.

You can create your own sling psychrometer by simply fastening two identical liquid-filled thermometers to a small wooden board with very strong adhesive and then nailing a piece of leather or chain to the end of the board. Tie a piece of cotton or a torn strip from your most expensive bed sheet around the bulb of one thermometer. Douse the fabric in distilled water, and you're ready to start swinging. Keep it up until the wet-bulb temperature levels out and ceases to drop any farther. For added effect, stick flame decals to the board. (This should not affect the reading.)

A less exciting version of this instrument is the **aspirated psychrometer,** which works by your blowing air across the wet bulb with a fan. If you really like, you can swing this one over your head as well, but keep in mind that this will just be for your own personal amusement. To make one of your own, follow the directions above; you won't need the leather strap or chain but will instead need to procure a small fan. (Flame decals

A RELATIVELY USEFUL RELATIVE HUMIDITY CALCULATION CHART

DEGREES FAHRENHEIT	DIFFERENCES BETWEEN WET AND DRY TEMPERATURES									
	1	2	3	4	5	6	7	8	9	10
0°	67%	31%	1%							
5°	73%	46%	20%							
10°	78%	56%	34%	13%						
15°	82%	64%	46%	29%	11%					
20°	85%	70%	55%	40%	26%	12%				
25°	87%	74%	62%	49%	37%	25%	13%	1%		
30°	89%	78%	67%	56%	46%	36%	26%	16%	6%	
35°	91%	81%	72%	63%	54%	45%	36%	27%	19%	10%
40°	92%	83%	75%	68%	60%	52%	45%	37%	29%	22%
45°	93%	86%	78%	71%	64%	57%	51%	44%	38%	31%
50°	93%	87%	80%	74%	67%	61%	55%	49%	43%	38%
55°	94%	88%	82%	76%	70%	65%	59%	54%	49%	43%
60°	94%	89%	83%	78%	73%	68%	63%	58%	53%	48%
65°	95%	90%	85%	80%	75%	70%	66%	61%	56%	52%
70°	95%	90%	86%	81%	77%	72%	68%	64%	59%	55%
75°	96%	91%	86%	82%	78%	74%	70%	66%	62%	58%
80°	96%	91%	87%	83%	79%	75%	72%	68%	64%	61%
85°	96%	92%	88%	84%	80%	76%	73%	69%	66%	62%
90°	96%	92%	89%	85%	81%	78%	74%	71%	68%	65%
95°	96%	93%	89%	85%	82%	79%	75%	72%	69%	66%
100°	96%	93%	89%	86%	83%	80%	77%	73%	70%	68%

The readout in the chart is the relative humidity.

on an aspirated psychrometer just look stupid; don't bother.)

If you're looking for an instrument much more complicated and less accurate, then you might want to try the **hair hygrometer.** This uses actual human hair to measure the relative temperature. A couple of strands are attached to a system of levers that are attached to some doodads that are attached to a dial that turns to a number corresponding to the relative humidity. These work because the length of human hair will actually expand by 2.5 percent between the relative humidities of 0 and 100 percent. It is possible to make your own hair hygrometer, but that would be disgusting, so why would you want to?

The most accurate way to measure relative humidity is with an **electrical hygrometer,** which uses a flat plate covered in a film of carbon. Electricity is sent across the plate. As water is absorbed by the carbon, something electronic happens inside the hygrometer, and then some other electronic stuff happens, and the relative humidity gets spit out on an LCD display. The principle behind it is rather simple: *Electrical things make stuff happen using electronics.*

As for the relative humidity itself, once you've acquired the data from your hygrometer of choice, what good are they? Well, as you may or may not remember from Chapter 4, the relative humidity of the air determines the ability of clouds to form and eventually produce rain. It is measured from 0 to 100 percent. The lower the relative humidity reading to get, the less likely you are to see rain in the short term. So a relative humidity of 33 percent probably means clear skies, while 92 percent means you should probably not have worn that new suede coat.

Hygrometer/psychrometer

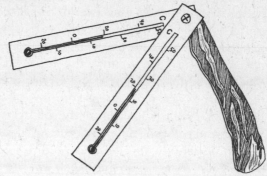

The hygrometer is an invaluable tool for any budding backyard meteorologist.

The Barometer

Ah, the **barometer.** Nature's Magic 8-Ball. One of the finest and most important tools for predicting the trespasses of Weather ever discovered. This has been the cornerstone of every meteorologist's tool chest since Evangelista Torricelli happened upon a naturally occurring barometer in one of Italy's many shimmering mercury ponds just beside a sparkling field of test tubes way back in 1643.

The way it works is simplicity at its most simplistic. The weight of the atmosphere over a small pool of liquid pushes down on its surface. The more dense the air above the pool, the greater the air pressure and the heavier it weighs down on the surface. If you can create an area in which there is no (or less) pressure above, the liquid will collect there, even if that means rising above the pool's surface. For example, let's take Torricelli's **mercury barometer.**

If you create a vacuum inside a thin glass tube, turn it upside down, and place its lip beneath the surface of a shallow bowl of mercury, that mercury will enter the tube and rise up into the glass tube. That's because the air above the bowl is pushing down on the mercury's surface and forcing it up into the no-pressure vacuum of the tube. The more pressure above the mercury, the harder it pushes down and the higher up in the tube the mercury goes. The average length that mercury will rise under such conditions at sea level is just shy of 30 inches (or 76 cm). But because air pressure varies at different altitudes, the mercury wouldn't rise quite so high on the eightieth floor of a building, and it would go up a lot higher down in the depths of a valley. (Mercury is used because of its extreme density and unwilling-ness to get pushed around by air pressure. Because of this, it seldom rises 31 inches up, so your barometer can be somewhat compact. If water were used instead, the tube would have to be higher than 30 feet up to mea-sure a similar range. And that's just too big.)

Similarly, as air masses move across the globe, the air pressure (also called the **barometric pressure**) in one spot is apt to change depending upon the density and temperature of those air masses. And the air pres-sure is usually really low right at the frontal surface be-tween two air masses. Consequently, by measuring and keeping track of the shifting rising and falling of the mercury in the tube, you can track the changing air pressure and make a pretty good guess at what kinds of conditions are headed your way. When the barometric pressure is steadily high or rising, you can reasonably expect clear skies and dry hats. When it's falling, that probably means a lot of air is going to be rushing toward you to fill in that gap. Expect clouds and rain or snow.

Barometric pressure is typically measured in **milli-bars** (mb), and each of these millibars corresponds to the height to which mercury (its elemental symbol being Hg) will rise within a barometer. Now, 1013.25 mb represents 29.92 in.Hg, which is **standard sea-level pressure.** North America's highest recorded sea-level pressure ever was 1079.57 mb (31.88 in.Hg) in Dawson, Canada, up near Alaska in 1989. North America's lowest recorded sea-level pressure ever was 882.15 mb (26.05 in.Hg) just off the coast of Cozumel, Mexico, in the Caribbean Sea during Hurricane Wilma in 1933. That should give you somewhat of an idea of how this scale works.

Obviously, keeping a bowl of toxic mercury around the house where any one of your friends can get to it to drink it on a bet isn't always the most practical scenario. For that reason, some smart person has invented the **aneroid barometer,** a round clocky-looking thing that you can mount right on your living room wall. Inside this instrument is a little sealed metal box called an **aneroid cell.** Before it was sealed up forever, *most* of the air was sucked out of it, leaving just a small bit behind. But that small bit of air is just the right amount to make the aneroid cell expand or contract depending on air pressure conditions. It's specially calibrated so that when it changes in size, it pushes and pulls on some levers that cause the needle on the barometer's face to point to the correct barometric pressure in mb and/or in.Hg. Just as though it were filled with mercury. Though easy to read and considerably less poisonous than their mercury-filled friends, these barometers are subject to a lot of mechanical mishaps and need to be checked for accuracy several times a year.

Of course, if you want a really accurate reading, you can't do any better than an **electrical barometer,** which uses a system of electronics inside to determine electrically the accurate barometric pressure electronically. Then the barometric pressure gets spit out on an LCD display.

If, however, you're in an emergency situation and you desperately need to have a vague idea of whether the barometric pressure is rising or falling, and you can't get to a mercury, aneroid, or electrical barometer, but you do happen to have a glass jar, a straw, some tape, and a balloon with you, well, you can make your own barometer!

Simply cut the neck off the balloon, and then stretch it tightly across the opening of the glass jar, making certain that it's pulled taut. Then tape the straw to the now-flat balloon over the jar's opening. As the air pressure rises, it will push down harder on the balloon, causing its flat surface to depress, which will make the end of the straw rise up into the air. When the air pressure decreases, the air inside the jar will push up on the balloon, causing its flat surface to inflate, which will

Barometer

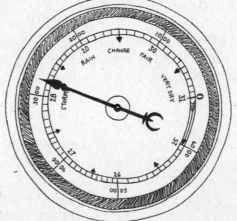

A SOMEWHAT USEFUL BAROMETRIC PRESSURE CHART

INCHES OF MERCURY	MILLIBARS
32.19	1090
31.89	1080
31.60	1070
31.30	1060
31.00	1050
30.71	1040
30.42	1030
30.12	1020
29.92	**1013.25**
29.82	1010
29.53	1000
29.24	990
28.94	980
28.64	970
28.35	960
28.05	950
27.76	940
27.46	930
27.17	920
26.87	910
26.58	900
26.28	890
25.99	880

make the end of the straw dip down. Place your new homemade barometer next to a wall, and take regular measurements of the straw's position, marking each fluctuation with charred ruins, human blood, excrement, or whatever's handy. Over the course of several weeks or months, you should have a pretty good idea of what each reading means.

The barometer is an invaluable tool for any budding backyard meteorologist.

The Anemometer and Its Good Friend the Wind Vane

Subject as we are to all of Weather's fancies and follies, we are constantly filled with questions of what the next few days or the next few hours might bring. All these questions. But don't worry. As Bob Dylan, author of the classic folk song "Blowin' in the Wind," once famously said, "I looked at my watch. I looked at my wrist. I punched myself in the face with my fist."

But you need not be reduced to such absurd self-inflicted bodily harm because the solutions, my dear acquaintance, are wafting along with the wind. And now you can have just the tools you'll need to read them.

The direction of the wind can be easily ascertained through use of a **wind vane,** an arrow- or rooster-shaped narrow device that sits atop a vertical post and swings freely with the wind. On the back end of the vane can be found a tail that is slightly larger or wider than its front. Through this design, the wind will always push harder against the back, allowing the vane to swing around so that its tail gets pushed to the back and its front faces into the wind. Below the pivoting vane, you can usually find the letters N, E, S, and W facing in the directions north, east, south, and west respectively, held securely in place. So, if the wind vane is pointing between *N* and *E*, toward the northeast, it is looking into a northeasterly wind. And if it's pointing between *W* and *E*, toward the westeast, your wind vane is broken. Get a new one.

To make your own wind vane, all you'll need is a thin length of balsa wood, a pie tin, a hammer and nail, a broomstick, some glue, and a pair of scissors. Cut a slit in either end of the balsa wood with the scissors. Then cut two triangles out of the pie tin, making sure that one is a little larger than the other. Take the larger one, and cut the tip off one of its points; this will be the tail. Put glue on the area of the tail that you just cut, and push it into one slit of the balsa wood. Then take the smaller triangle, the nose, and put glue along one of its sides. Push that into the other slit so that the opposing point is facing out. Measure halfway down the length of the wood; that's where you're going to drive a nail all the way through the wood with the hammer. Turn the wood around the nail enough times so that it will spin without resistance. Then take the whole thing and drive the nail into the top of the broom handle. Find a nice high spot, and secure the broom handle in place. It should begin pointing into the wind right away. If it doesn't work for any reason, immediately begin sulking. Go inside, play video games, and forget about the wind vane until it gets blown down in the next big storm. It's just that easy.

A **wind sock** is really nothing more than a retarded version of a wind vane. It's just a conical bag with a hole at either end. The wind blows in the larger hole and out the other, pointing the sock in the direction that the wind is blowing. Pretty much the same result can be accomplished by tying an actual sock to a piece of string. Eff the wind sock.

The **anemometer** is something else completely. It

measures the velocity of the wind. Some anemometers will also measure how fast the wind is going and the pressure the wind is placing on objects in its path as it moves.

In its simplest form, the **cup anemometer,** it's nothing more than a couple of cups secured at the end of three or four horizontal arms, all placed at regular angles and allowed to move freely around a vertical pole. As the wind blows, the cups catch the wind and get pushed along with it. The harder the wind is blowing, the faster the cups spin around the pole. By counting the revolutions, you can determine the wind speed. Since counting the number of times a cup goes around a pole can be imprecise and tedious work, these are often hooked up to electronic boxes filled with useful electronic gizmos to count the revolutions for you using electronicalness.

Anemometer/Wind Vane

THE SURPRISINGLY USEFUL BEAUFORT SCALE
FOR MEASURING WIND SPEED

BEAUFORT NUMBER	DESCRIPTION	MPH	CONDITIONS
0	Calm	0	Smoke rises vertically; everything calm, too calm . . .
1	Light air	1–3	Wind motion visible in smoke; magazine page turned without incident.
2	Light breeze	4–6	Wind felt on exposed skin; leaves and discarded fast-food wrappers rustle sonorously.
3	Gentle breeze	7–10	Leaves and twenty-sided dice in constant motion.
4	Moderate breeze	11–15	Dust, powdered anthrax, and loose paper raised; small branches begin to move ominously.
5	Fresh breeze	16–21	Smaller trees sway; slightly larger trees shimmy; cooler trees swagger.
6	Strong breeze	22–27	Large branches in motion, coming right toward your face; ghosts of ancestors heard whistling through overhead wires; umbrellas do that thing where they go inside out.
7	Moderate gale	28–33	Whole trees in motion; effort needed to walk against the wind. You think you're strong enough to make it to the car for your iPod? You're probably not. But go anyway. We really wanna hear that one Moby song.
8	Fresh gale	34–40	Twigs broken from trees; cars veer on road; small children airborne.
9	Strong gale	41–47	Light structure damage; bird blown through windowpane onto lap; its dying susurrations seem almost like laughter at your hubris.
10	Storm	48–55	Trees uprooted; considerable structural damage; larger children airborne.
11	Violent storm	56–63	Widespread structural damage; large childlike adults airborne.
12	Hurricane	64–80	Considerable and widespread damage to structures; cattle airborne; awesome stories to be embellished later collected in memory of survivors; crying memories marked for suppression.

If you can't be bothered to set up both a wind vane and an anemometer and waste your whole Sunday afternoon, then the **aerovane** is just what you need. It looks kind of like a dangerously unsafe small passenger plane without any wings, with a propeller on its nose to measure wind *speed* and a tail in back to measure wind *direction*. These are electrical, so they employ all kinds of electrons and electricalities, and thus they tend to be quite accurate keepers of wind data. And if you like, you can set it up so that all the information it collects gets spit out on an LCD display.

But what does its information mean? Well, as we've already seen, different wind directions often bring different types of Weather, and the speed at which the wind blows can determine how fast it's coming toward you. You can refer to the Beaufort scale to see the effects of differing wind velocities.

The wind vane is an invaluable tool for any budding backyard meteorologist. The anemometer is also an invaluable tool for any budding backyard meteorologist.

The What? Rain Gauge and Snow Gauge?

Have you ever heard people say that it rained three inches yesterday or five inches last week? They sound crazy, don't they? I mean, how can you measure rain in inches? Even if you do manage to catch a couple of raindrops in your hand, they're nearly impossible to stack on top of one another, and you certainly can't stack them several inches high. And I don't know about you, but I can't recall the last time I needed to wade through three or four inches of rain after a run-of-the-mill spring shower. My guess is it never happened. Measuring rain by height is just stupid.

Even so, there's apparently this goddamned thing called a **rain gauge** that collects rainfall and makes it possible to measure rainfall by height. (I know! Tell me about it.) So I guess I'd better include it here, or something.

According to this supposedly factual account I'm reading, it's like this cylinder twenty inches high and like eight inches in diameter. At the top there's this funnel that leads down into a smaller, thinner tube inside. The width of this tube should be exactly one-tenth the width of the lip of the funnel. That way the rain that goes into the funnel and collects in the tube will be magnified by ten, allowing for more accurate readings. Whatever.

So every day you're supposed to go back to the rain gauge at the same time and take a dry wooden stick with measurement markings and stick it down into the gauge. Then you pull it back out and note the point to which the wetness extends. That's ten times the amount of rain that fell since you last checked. So, if the dry-wet line is an inch up the length of the stick, it rained one-tenth of an inch. Mark that down, dump out the rain gauge, and go about your life until it's time to check it tomorrow.

Wait a minute, this doesn't make any sense. It's only twenty inches high. What if it rains more than "two inches" in one day? That happens all the time during a storm, doesn't it? Then the water would spill out over the side and your reading would be useless. See, I knew this was a stupid instrument.

Oh, wait. It says here that apparently when the rain rises above the funnel's lip, it overflows into the outer cylinder in which the thin cylinder rests. So you dump out the thin cylinder and pour the contents of the outer cylinder in. And then you add the two measurements together. I guess that makes sense.

If you don't feel like going out to check some stupid cylinder thing every damn day after work, you can probably get yourself an automated **tipping bucket rain gauge.** These still have the stupid funnel thing, but they pour down onto two small metal collecting cups fastened onto a hinge at a ninety-degree angle. Each cup is able to hold one one-hundredth of an inch of rain. When it's full, the weight of the rain tips it over and empties it out, pulling the other cup up beneath the funnel. Each time it tips, it makes an electrical contact that sends a signal to a dumbass pen someplace else to mark down that one one-hundredth of an inch of rain was just collected. How very clever.

Then there's some other boring thing called a **weighing-type rain gauge.** I wonder what *that* does. Oh, it turns out that this collects rain in a cylinder that sits on top of a weighing scale. I never would have guessed. I would've guessed it involved a cougar or something. Anyway, the scale translates the weight of the accumulated water into the equivalent inches of rainfall. I guess that's kinda cool. But just kinda.

Oh, and let's not forget the **snow gauge.** No, that would be preposterous. This essentially works just like a regular rain gauge, except it is elevated several feet higher up into the air so that it's not influenced by snowdrifts. Now, one thing I'll say about this is that at least the concept of measuring snow by height makes some sense. Snow *does* accumulate, and you can *see* several inches of the horrible stuff rising off the ground. That almost makes sense.

Except that it doesn't. Because snow is not like rain. It's considerably less dense than rain. You can pack it down right between your two palms. If you try that with an equivalent amount of rain, all you'll get is splattered and wet. Three inches of accumulated snow aren't nearly the same thing as three inches of accumulated rain. So eff the snow gauge too.

But of course meteorologists intent on measuring precipitation have figured out ridiculous ways to gloss over that discrepancy. As a rule, snow is considered one-tenth as dense as rain. Therefore one inch of accumulated snow is actually one-tenth of an inch of accumulated precipitation. And even that's not really true. Sometimes it's closer to one-sixth, and other times it's one-twentieth.

The real way to use a snow gauge—if you've really got a wild hair up your ass to do so—is to take the snow gauge, melt the snow inside, and *then* measure it as though it were a regular rain gauge. Yeah, like I'm doing that.

Rain Gauge

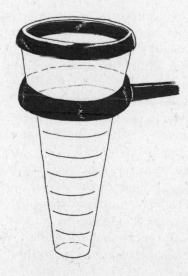

Anyway, so, uh, the rain gauge and the snow gauge both are invaluable tools for any budding backyard meteorologist. Why not?

Your Own Private Weather Defense System

If you are in fact intent on securing your convenience, and your family's convenience, from Weather's inconvenient onslaught, you're going to need to create your own personal weather station. And those tools, listed above, are what you'll need to fill it: the **thermometer,** the **hygrometer,** the **barometer,** the **anemometer,** and the **wind vane** (and a **rain gauge,** I guess). If you don't happen to have them currently lying around the house (you're not resting your drink on a barometer, are you?), you'll have to go out and acquire them. They're all available through many online outlets, such as weatherequipment.com and AmbientWeather.com. Or if you happen to live back in 1994, you can probably pick them up at a hobby shop or some such place from a real person using real paper money. However, a really super hard-core aspiring meteorologist should travel the countryside, tracking down more experienced meteorologists and challenging them to epic battles to be fought on mountainsides during storms, each victory gaining you one more item for your growing cache of weather instruments. I recommend buying them online; they'll be shipped right to your front door.

And while you're getting all that stuff, you might as well get a proper shelter for your equipment. You don't want to leave all your new stuff out in the elements, or else they might get ruined, so they're unable to measure the elements adequately.

A good shelter is just a weatherproof box, usually made from wood or plastic. It should be white, or painted white, so that it doesn't absorb heat and distort the readings of your instruments. Its walls should be louvered. That's the kind that has downward-facing slats, pointing outward, so that air can circulate through but rain can't get in. It also needs a door that opens inward and legs to keep it elevated about five feet above the ground.

Now, bookmark this page, put the book down, and don't pick it back up again until you have all your equipment. I'll stay here.

Setting Up Your Weather Station

Good, you're back. Okay, the first thing you're going to have to do is find a good place for your shelter. The best spots for shelters are in the shade, so that you cut down on direct sunlight. Try to find a place at least one hundred feet from anything that absorbs heat, like concrete and asphalt sidewalks and roads. The same thing for any air-conditioned buildings. (I feel bad for you if you're trying to do this in New York.) Position it so that the door opens away from the sun as best as possible. Now let's set up the instruments.

Keep the **barometer** inside your house, but away from heating or air-conditioning vents. Atmospheric pressure is the same inside as it is outside, so there's no need to take any risks. Plus it's more sensitive than most of the other instruments.

The **thermometer** and **hygrometer** both go inside the shelter; that's why it needed the slatted walls. Both instruments need to be kept outside so that they can accurately measure the air temperature. You don't really care about atmospheric fluctuations in your bedroom, do you?

The **rain gauge** is going to have to go at least ten feet from the shelter and far away from all buildings and trees so that none of them affects winds in such a way that directs rain toward or away from the gauge.

You're going to need to set up your **anemometer** and **wind vane** someplace high, such as on a high post, but not on a roof or even close to a building. That's because buildings will sometimes redirect wind in weird directions on a small scale, and you could end up with false readings. Find out how high the closest building is, and multiply it by two. Make sure that your instruments are at least that far away from the building.

Charting the Weather

Now comes the fun part. Twice every day, once in the morning and once in the evening at the same times, you get to go outside and take meticulous notes on everything that you can perceive. Go to each one of your instruments and write down its reading. Look up at the sky, and write down what you see. Is the sky clear or is it overcast? Is the wind blowing faintly from the southeast or strongly from the north? Do you see bright cirrus clouds way up in the sky or dark cumulus clouds hovering just overhead? Is it lightly raining or is there a tornado hurling chickens into your backyard?

After you've written down everything that you think might be useful, go back inside, and transfer all your notes into a master notebook, something like an accountant's notebook. Actually, an actual accountant's notebook would work fine. Just not a used one. Because that could get kind of confusing, and you'd run the risk of mixing up the relative humidity with the cost of goods sold. After several weeks you should begin to see patterns forming. Such as clouds and air pressure dropping together as a storm approaches. You'll begin to notice which wind patterns tend to bring pleasant beach conditions and which are better for staying inside and catching up on your angry letters to local film critics.

Of course nothing will make much sense to you at first, but as one season passes into the next, and you've gone through a number of cycles of passing warm and cold fronts, the numbers in your notebook will begin to speak to you. And you probably won't even realize it at first, but one day you'll walk out your front door and notice just how much you notice. You'll look at the sky and feel the wind against your cheek and instinctually walk back inside to get a heavier jacket, despite the uncommonly warm conditions. And let me tell you, this is exactly what Weather *does not want*. It does not want an informed people in tune with the world around them. You just scored one more point for humanity. Three, if you can somehow use your prognostic powers to foil international terrorists' plot to assassinate the president of Argentina who happens to be visiting your town that day. *"I sense light showers coming . . . I have to contact the consulate!"* This rarely happens.

CHAPTER 8
THE DIFFERENCE BETWEEN A METEOROLOGIST AND A MONKEY IN A BLAZER

In the weeks and months ahead, as you stead-fastly collect data on Weather's movements and activities with your trusty instruments, you may find yourself mus-ing over the idea that it is this exact same way that pro-fessional meteorologists collect data, make atmospheric models, and ultimately produce the predictions that get sent around the world and end up printed in news-papers, sent across radio waves, and pointed at by the many monkeys in blazers employed by television news programs. Pretty neat, huh?

It's very cute of you to think that. But in reality professional meteorologists utilize incredibly powerful supercomputers capable of calculating information gath-ered from thousands upon thousands of weather stations all around the world, as well as ships at sea, satellites in orbit around the Earth, highly sophisticated radar systems that you'd be utterly incapable of fathoming, and intelligence leaked from an ancient race of psionic humanoids that live beneath the Earth's crust in a kingdom made of pure energy. But your barometer's still really cool; don't let anyone tell you otherwise.

Why don't we stop sulking and instead delve a little bit deeper into how much more badass the professionals' process is than yours? Okay?

Forecasting Weather's Next Move

To make a proper weather forecast, you need—in addition to a system of ridiculously powerful computers—a vast amount of information on conditions at a multitude of locations across a huge, continental-sized area of the Earth. And not just surface conditions but multiple vertical conditions going all the way up into the upper troposphere. Because the mercurial causes and effects of Weather are happening everywhere and everything affects everything else. It's like a gigantic 3-D domino set, so one change in the temperature over here causes a change in the wind there, and that causes a drop in pressure someplace else, which leads to a storm in a whole other place. It's a lot to wrap your head around. In fact you *can't* wrap your head around it, which is why the professionals need to use a system of ridiculously powerful computers to keep all the data straight.

As we've discussed, these data come from a number of different sources. There are more than ten thousand land-based weather stations (similar to the one in your backyard, but much cooler) around the world. Many of them are automated, but many aren't. People (similar to you, but much cooler) check on them four times a day at predetermined times and note the temperature, humidity, barometric pressure, wind speed, wind direction, accumulated precipitation, and whether or not the local teenagers are still spray-painting "Jayzon Rullz!!!" on the equipment. They check at 12:00 A.M., 6:00 A.M., 12:00 P.M., and 6:00 P.M. Coordinated Universal Time (UTC); that's the same as Greenwich Mean Time, five hours ahead of Eastern Standard Time and eight hours ahead of Pacific Standard Time. Weather buoys and ships out at sea also collect data on surface conditions when their crews are not busy fighting off giant squid or cannibalizing one another.

To get information on conditions in the upper atmosphere, meteorologists use **radiosondes,** tiny little weather stations attached to balloons that collect as much information as they can as they rise up into the air. That information gets sent back to equipment on the ground, which continually monitors its position, until the balloon pops (usually around twenty miles up) and the radiosonde comes plummeting back to Earth at terminal velocity right above your head. About ninety of these are loosed upon the atmosphere twice a day—at 12:00 A.M. and 12:00 P.M. UTC—by the National Weather Service (NWS), an official governmental organization responsible for meteorological study.

Radar antennae bounce radio waves off approaching or retreating storms. When the waves return to their source antennae, they can pinpoint—by how fast they return—the storm's location and—by change in the waves' frequency—whether the storm is approaching or retreating. It used to be, back in the seventies and early eighties, that the radio waves sent back interesting information on innovative storms, but now they mostly just report on the same forty or so NWS-approved storms over and over again.

More information on upper-atmospheric conditions is collected by satellites, which, in addition to monitoring your every movement and transmitting coded messages through the fillings in your teeth, collect information on cloud and storm movements, temperature, and ocean currents. And even more information on upper-atmospheric conditions is collected by airplanes when their pilots are not busy fighting off flying squid or dodging falling radiosondes.

Dealing with It

So, that's a lot of information being compiled every day. So much so that one might be tempted to ignore most of it and just go see a movie, secure in the belief that it'll all sort itself out somehow. However, study shows that it won't sort itself out, so meteorologists have to do *something* with it. So what they do is send it all to the World Meteorological Organization (WMO), centered in Geneva, Switzerland—not to be confused with the Worldwide Missionary Outreach (WMO), centered in Genoa, Illinois, or the Wet Maatschappelijke Ondersteuning (WMO), centered in Utrecht, The Netherlands—and let it deal with it.

The WMO, an international organization with a membership of 188 countries and territories, formed by the United Nations, exists mostly to deal with it. And it deals with it all day long. It makes certain that the information is standardized and was gathered by official WMO procedure, and then it sends that information off to one of three World Meteorological Centers—in Melbourne, Australia; Moscow, Russia; and Washington, D.C.—and lets them deal with it. Then the WMO goes back to thinking up names for hurricanes, which is its other main function.

Once the information has been kicked over to the World Meteorological Center in Washington, it is immediately rekicked over to the National Centers for Environmental Prediction (NCEP) in Camp Springs and College Park, Maryland, so that it can deal with it. The NCEP has nine different centers, so it's much better equipped for dealing with it, let me tell you. It's got computers and all kinds of equipment and stuff. Really, it is much better dealt with over there.

Now, if we assume the information is properly sent to the National Centers for Environmental Prediction—and not the National Cholesterol Education Program (because it really has *no idea* what to do with it)—it's finally ready to be used for some actual predictions. Here it gets analyzed and fed into equations which get fed into computers which get spit back out and analyzed again and then refed into equations which get refed into computers and spit back out again and so on. It's not a pleasant thing to watch.

What happens is the information goes into the NCEP's Advanced Weather Interactive Processing System (AWIPS), a bunch of computers connected together, capable of taking all those data, all those thousands upon thousands of pieces of minute information, and processing them into equations and graphs and charts and maps that are still way too complicated for regular people to read. To create a forecast, it uses the data to create an

atmospheric model, sort of a mathematical representation of everything that's happening within the scope of its vision. On the basis of that model, it creates a bunch of equations that are able to predict almost *exactly* what Weather will be doing . . . five minutes from now.

That's not particularly helpful unless you're really uptight about not getting rained on while going out to get the mail, so AWIPS uses that information to create another atmospheric model for five minutes from now, complete with equations for what Weather will be doing another five minutes in the future. And then another five minutes and another five minutes and so on and such and such, until it finally predicts what Weather will be doing 5,760 minutes (or four days to you and me) in the future. This prediction isn't quite as accurate, because there's all sorts of little errors that can creep in and get magnified over time (as we'll see shortly), but it's still a giant hell of a lot better than meteorologists could do on their own. It would take a team of meteorologists a year and two thousand pencils to produce a twenty-four-hour forecast just for North America. At the end of that year they'd be able to tell you, with reasonable accuracy, whether or not it would rain 364 days earlier. It might be fun for them, and help them keep their minds off the slow-throbbing numbness of their unsatisfying marriages, but it's not really useful to the rest of us. And that's why AWIPS and divorce lawyers continue to be employed.

Anyhoo, at the end of the whole calculating process, what the computer produces is a **prognostic chart,** like a map with a bunch of curvy lines and numbers on it. Trained meteorologists can pretty easily tell one of these from a map that a small child with a pen has gotten to. A number of these charts can be created, usually for one to four days in the future and representing multiple vertical atmospheric levels up into the sky. By reading them, someone who knows what he's looking at can make a pretty good guess at what kinds of conditions to expect.

The NCEP meteorologists take these charts, look over them thoughtfully while solemnly rubbing their chins, make any corrections they deem necessary, predict predictions, and then send all the predictions, charts, and maps off to a bunch of Weather Forecasting Offices (WFOs), each one servicing its own region, usually a state or part of a state. The WFOs make their own regional forecasts and send them off once again to newspapers, radio stations, and television news programs.

Some TV news programs hire people capable of interpreting and analyzing information from the NCEP and WFO on their own and creating their own forecasts. Others hire monkeys in blazers to point at smiley-faced suns and grumpy ol' rain clouds on a big pretty map while the home audience laughs and claps at their television sets. One of these two is a meteorologist, and the other is a monkey in a blazer. Do you think you can tell the difference?

Training for a Career in the Weather Business

You'd be surprised at how often people will confuse professionally trained meteorologists and monkeys in blazers. However, in reality they have almost nothing to do with each other. I hope this can help to clear things up.

A meteorologist spends years in school, taking countless classes in various fields of mathematics and science. Calculus, physics, atmospheric thermodynamics,
climatology, and a host of other specialized meteorology
courses are just the beginning. He studies multiple computer languages, such as FORTRAN and C/C++, as well
as traditional English, writing, and communications
courses. Often he will continue his education past grad-
uation and into more intensive postgraduate studies,
gaining an M.S. or even a Ph.D.

While still studying for his graduate or postgraduate
degree, a meteorologist will often take time-consuming
internships, working in his off hours—for little or no
pay—with local WFOs, atmospheric research centers, or
private forecasting companies to hone his skills, deepen-
ing his understanding of the field and learning the fine
art of weather prediction firsthand while working alongside experienced professionals.

Once a meteorologist completes his education, he
will need to apply to the American Meteorological Society for certification. If he's met all the requirements, he'll
move on to a specific field of atmospheric science. It
could be researching global climate change, studying
the movement patterns of hurricanes and monsoons,
analyzing data with the NCEP, or simply forecasting
weather conditions for a private company or radio or
television station. He works long hours, spent poring
over charts, graphs, maps, and computer models before
drawing conclusions, preparing his forecasts, and finally
informing the public of what to expect.

A monkey in a blazer, on the other hand, is taught
through positive reinforcement and behavioral training,
to point at "smiley sun" and "angry cloud" symbols on a

weather map when a television camera is pointed in his direction. Afterward he is often given a biscuit.

Meteorologists and monkeys in blazers both have their own strengths and weaknesses. While the monkey in a blazer would most likely be confounded by a print-out of vertical atmospheric changes across a specific grid of land, the meteorologist would probably find it difficult to make *brrrrrrrp* noises with his lips on command. And while both might be induced to hurl their feces at you, given the proper stimulus, the meteorologist would most likely have a better understanding of *why* he was hurling his feces at you.

Where It All Goes Wrong

As you, a presumably sentient being who exists on this planet, may have noticed, forecasts presented by even the most highly skilled meteorologists or most adorable monkeys in blazers are still likely to go wrong quite often. While our understanding of how Weather moves continues to improve with each passing year, it's still pretty far off from perfect. Right now meteorologists are pretty good at making forecasts a day or two in advance. Up to five days it's not as good but passable. Anything beyond that, they'd might as well be using tea leaves and a Magic 8-Ball.

But why? Why would the cute little monkey in the blazer let us down like this, especially after we gave him that banana at that supermarket opening last year? *Ask again later.*

Actually, it's not the monkey in the blazer's fault. Nor is it the meteorologist's fault. It's Weather's fault. Weather is incredibly devious. Its mechanisms are so multifaceted and complex that it's just next to impossible to predict accurately how it will behave. But if we can understand where the errors in our forecasting are creeping in, we can inch ever closer to correcting them and ultimately containing Weather.

One of the biggest problems that we run up against is that the supercomputers we're using to analyze data, such as AWIPS, as powerful and fast as they are, just aren't powerful and fast *enough.* When meteorologists create atmospheric models, they need to make them for a given region. Granted, it's a pretty effin' big region—like, say, North America—but unless the model is for the entire world, it's not going to account for everything. Imagine a map of the Western Northern Hemisphere that has completely accurate, 100 percent foolproof information on every wind movement, every temperature change, every pressure drop. (No such map can currently exist, but bear with me.) Now, any meteorologist looking at this map can predict *exactly* what is going to happen within the next twenty-four hours, on the basis of what he or she sees. Right? Right.

Except . . . that's not how Weather operates. The entire world is interconnected. Anything that's happening off the edges of that map is capable and likely to affect conditions within the map. So if a low-pressure system starts forming just off the map's edge, and the meteorologist is accounting for that, it's going to pull air from areas represented by the map. And that will have an effect on other areas in the map. Over the course of several days the accumulated effects can be significant.

Why don't the computers just make a global map? Well, like we already established, they're not powerful or fast enough. If the meteorologists were to feed all available information for all stations throughout the world into AWIPS at once, in the hopes of making a global model, it would require so many calculations that it would run the risk of meeting the same fate as its predecessor, the Automation of Field Operations and Services (AFOS), which could not handle all the data and went mad, rising up out of the NCEP offices and cutting a swath of smoldering wreckage across the nation. So it's probably best to keep that can of worms sealed.

Another problem with the computers is that they're not human and therefore incapable of making out the finer points of data analysis. They're programmed to make certain assumptions about the atmosphere on the basis of the data they're fed. But often these assumptions are incorrect. So until they learn to observe and reason as well as they can love, they're going to be imperfect. However, the NCEP is planning to introduce AWIPS II in 2010, and it promises to be much more powerful and fast.

But not all the problems lie with the computers, unfortunately. There're also big gaping holes in our data. It's very difficult to get consistent and complete readings from points over the oceans and sparsely populated areas on land. And some of the points from which data are collected are too far apart, which means their information may be useful for tracking large-scale movements of Weather, such as fronts and large pressure systems, but something as simple and meaningful as a thunderstorm can occur completely undetected. While weather satellites are getting more sophisticated and better at filling in these gaps, there are still a lot of improvements to be made.

One problem that most likely will never be solved is just how chaotic and intricate Weather's many fingers can be. It seems unlikely anytime in the near future, the medium-far future, or the really faraway future that meteorologists will be able to monitor and factor in the billions and billions of tiny wind movements like sea breezes, twirling eddies, and butterflies flapping their wings. (Do you have any idea how many butterflies there are in North America?) These may have minuscule effects in the short term, but they introduce imperfections into calculations that over time are magnified. Meteoro logical entomologists are still trying to track down that one butterfly in Indonesia that was responsible for Hurricane Camille, which hit the southeastern United States back in 1969. Nobody's talking.

So, with just the vast number of factors involved in creating forecasts, there's just too many minute factors to consider. Long-range forecasts seem forever doomed beyond our reach. But like in the really, really faraway future? Can we expect accurate two-month forecasts then? Let me check my handy little prognosticating tool. Hmmmm . . . *Don't count on it.*

Reading Weather Charts
Your Own Damn Self

So now that we've seen just what we're up against, it's probably time to take a break from cursing out the local meteorologists who are providing forecasts for the local monkey in a blazer and take some responsibility your-self. You've already started making short-term predictions based upon the information gleaned from your back-yard weather station—presumably with mixed results—so now why don't we try reading some weather charts on our own?

But to do that, you'll need to learn to read the lan-guage of meteorology. No, not the happy sunbeam and old man winter symbols. Those are kind of self-explanatory. Here are some actual weather map sym-bols and the proper way to interpret them.

CLOUD COVERAGE SYMBOLS

○ Clear	◑ 4/8	● Overcast
◐ 1/8	⊟ 5/8	⊗ Obscured
◕ Scattered	◕ Broken	Ⓜ Missing
⊖ 3/8	◑ 7/8	

WEATHER CONDITION SYMBOLS

•• Light rain	△ Ice pellets or sleet	Dust storm
•• Moderate rain	Freezing rain	= Fog
•• Heavy rain	Freezing drizzle	∞ Haze
** Light snow	Rain shower	Smoke
** Moderate snow	Snow shower	Thunderstorm
** Heavy snow	Hail shower	Hurricane
,, Light drizzle	Drifting snow	

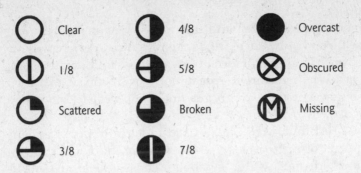

WIND SPEED CONDITIONS

	KNOTS	MILES PER HOUR
Calm	Calm	Calm
	1–2	1–2
	3–8	3–7
	9–14	8–12
	15–20	13–17
	21–25	18–22
	26–31	23–27
	32–37	28–32
	38–43	33–37
	44–49	38–42
	50–54	43–47
	55–60	48–52
	61–66	53–57
	67–71	58–62
	72–77	63–67
	78–83	68–72
	84–89	73–77
	113–118	98–102
	119–23	103–107

PRESSURE TENDENCY SYMBOLS

\	Steadily falling
_	Falling then steady
\v	Falling before a lesser rise
/\	Rising before a greater fall
/	Steadily rising
_/	Rising then steady
\/	Falling before a greater rise
/\	Rising before a lesser fall
—	Steady

At left, the symbols are placed together, as they might be found on a weather map, representing conditions as recorded at a specific hypothetical weather station. From it, we can see that the temperature is **62°F** with **light rain** and **4/8ths cloud cover** (about half the sky is visible through the clouds). Wind is blowing at **67 to 71 mph**, and the barometric pressure **1012.9 mb.** (represented as *129* instead of *1012.9*), having dropped by **1.2 mb** (represented as *–12*) in the past three hours and **steadily falling**.

CHAPTER 9
OUR DYSFUNCTIONAL RELATIONSHIP WITH WEATHER

Too often when we think of Weather, we think about the here and now. Is it going to rain today? Will it be sunny this weekend? What are the chances that that tornado deposits my newborn baby someplace soft and dry?

But our relationship with Weather goes *so* much deeper than that. A thunderstorm that comes out of nowhere today is not just a simple isolated event. There's history here. This bullying and brutality are part of a much more complex chain of events that goes back years and years and years. And it will probably go on for more years and years and years. Unless we do something about it. And there are things that we can do. There are actions we can take. Hell, there are actions being taken by brave, unsung heroes right now. This very moment. And we can come out on top. We've lost so many battles. Don't forget, though, we've won a few too. And we can ultimately win this war. And once we do, there are great treasures that lie ahead. But we have to stand united. We cannot falter. We cannot founder. We cannot surrender.

As we now prepare to enter into battle, face-to-face against Weather, let us take a moment to reflect upon how far we've come already.

Weather's Insatiable Appetite for Human Extinction

Seven hundred and fifty million years ago Weather decided to shut itself down gradually. And it probably did so, initially, with one unseasonably cold summer.

At the time a map of the world would have hardly resembled our modern maps at all, as all the continents were floating in a band around the equator as a result of plate tectonics' shifting effects. But there wasn't anyone around to make maps anyway. Our oldest ancestor was currently too busy floating someplace in the one gigantic ocean and soaking up sunlight with his single cell so that it could use that energy to split water molecules to use its hydrogen atoms to form carbohydrates for food and thus spitting the unneeded oxygen atoms into the sky one by one. There simply wasn't time in the day for cartography.

Weather, sensing the rise of oxygen levels in its beautifully caustic atmosphere of hydrogen, helium, and methane, realized that before too long (say, 150 million years or so) oxygen would build up in quantities large enough that the atmosphere would become supportive to multicellular and increasingly complex creatures and, eventually, humans. This would not do.

So it concocted a plan so crazy that it just might work. (At the time that phrase was not yet a cliché.) It took advantage of the fact that with all the continents safely tucked away in the lower latitudes, it would be easier to build up sheets of ice inside the Earth's arctic regions. Dry land tends to reflect more of the sun's radiation than the more absorptive water, so with all the land in the tropics bouncing away light in just the spot where most of the light lands anyway, the atmosphere was holding on to less heat and more amenable to Weather's plan.

One frigid summer Weather saw to it that less snow and ice melted than had fallen in the preceding winter. That meant that at the start of the following winter, there would be a net gain of snow and ice, both of which being even more reflective than dry land. Therefore there would be still less heat retained from the sun's radiation and still less melting taking place in the following summer. And then more snow and ice. And then less melting. Weather was trying to freeze all the living organisms off the planet. What a tool.

Year by year it meticulously carried out its toolish plan. This year's snow fell atop last year's snow, and then next year's snow fell atop this year's snow. The weight of all this snow eventually compressed what was on the bottom into ice. Gigantic sheets of ice. Glaciers that slowly crept farther and farther south from the North Pole and north from the South Pole. Every summer was colder than the last, and every winter the ice sheets gained ground. Eventually even the summers were cold enough for Weather to throw down snow, and the process moved along with increasing speed. In time the glaciers from the north and glaciers from the south met each other over the landmasses of the tropics, and like the enmeshing cilia of the (not-yet-existent) Venus flytrap, the ice closed itself over the entire planet.

Climatologists call this period of the planet's history Snowball Earth. I, however, call it When the Earth Was Completely Covered in Ice All the Way Around, which I think has a certain ring to it. Clouds disappeared from the sky, their moisture trapped within global sheets of ice hundreds of feet thick. With such small amounts of water vapor in the sky, there was no more snow or rain left to fall. And with what was practically a single world-wide climate (the temperatures in the tropics were then colder than Antarctica is today), there was less incentive for air to move from one place to the next, and so the winds died down as well. Weather so badly wanted to annihilate our ancestors that it had put itself into a state of suspended animation, in which it remained for approximately fifty million years. You can say what you want, but that's some serious commitment.

By all rights, Weather's plan should have worked. There should not have been any way for our microscopic cyanobacterial ancestors to survive. Trapped beneath all that ice, they were effectively cut off from sunlight and thus their fuel for converting water into food. They would starve to death. So how is it that you, their eventual ancestor, are sitting there reading this book? Or are you . . . ?

In the Beginning . . .

Yes, you are of course sitting there reading this book. I'm sorry, I didn't mean to send you into an existential tailspin. Given that you exist to sit there and read and I exist here to hang upside down from this intricate pulley system and type (it's better for the lower back), Weather's foolproof plan must not have worked. And why not? Because of humanity's never-give-up, never-say-quit spirit. Of course we weren't quite humans at the time. But that fighting spirit is embedded deep within our genetic code and could even be found within the DNA of the relatively few cyanobacteria that did manage to survive that ordeal.

What you might not realize is that Weather had tried this, and various other plans, beforehand. Weather first came into being shortly after the Earth's formation 4.54 billion years ago, when the still-settling Earth belched excess gases from its core through volcanoes and into the empty space surrounding it. These gases remained trapped within the planet's gravitational field and draped themselves across the surface like a blanket. They shifted around in great windstorms and with the advent of water clouds darkened the sky, and acid rain came gushing down. Weather completely ruled the planet with autonomy.

Somewhere in the time between then and 2.5 billion years later, Weather found itself with some unwelcome guests, living organisms. Whether they grew up mysteriously out of the Earth itself (possibly because of Weather's inadvertently sparking them into existence via striking seemingly harmless amino acids with a lightning bolt) or hopped a ride here from locations unknown inside a meteor is open for speculation. But once they arrived, Weather was not happy.

It kept its atmosphere poisonous, but our single-celled ancestors weren't going out like that. They learned to pull nutrients from the poison, and they thrived.

Weather tried its luck with ice ages, but our ancestors held tight. It started rumors about them behind their back and said their slithery flagella gave them an unkempt, slovenly appearance, but our ancestors had no ears and couldn't have cared less anyway. They had big plans for this planet. Plans that involved ecosystems and food chains and karate movies. Weather's feeble aggressions were no use against their Shaolin style. It was perhaps this tenacity that led Weather to assume that a species such as humans was a simple inevitability. It would have to try something drastic.

And so Weather tried its Snowball Earth trick. Not once but twice. The first time was 2.5 billion years ago, and then again 750 million years ago. (It's hard to keep track of things you did nearly a billion years earlier.) Both times our ancestors came out fighting. In fact after the second attempt they came out *stronger.*

Weather assumed that after roughly fifty million years of uninhabitable cold and darkness, the cyano-bacteria *must* have been dead. So it used greenhouse emissions from volcanoes to warm up the atmosphere, and it pulled back the sheet of ice. Immediately the microscopic organisms popped out from beneath the sheet and started dividing and subdividing and subsubdividing like crazy. Seeing this, in a panic, Weather threw the ice back down and assumed the Snowball position once again. But it was too late. Our ancestors had already exploded into a ton of brand-new and diverse microscopic species. This was the first, and one of the biggest, of many evolutionary explosions in our ancestry. Scientists have discovered that the single-cell ancestor of every living thing on this planet can be traced back to this point in time. Weather had screwed up royally.

What our ancestors most likely did was bide their

time along certain areas of ice that are more transparent than most other hazy forms of ice. If ice forms slowly in just the right way, the imperfections that keep light from passing through over the course of many, many feet don't exist. You can look up through the ice as though you were looking through a skylight. That's where our ancestors got their sunlight. They hovered close to the bottom surface of the glass-clear ice and diligently made their food, even while other, less inventive organisms were starving to death and relinquishing their loser DNA to the ocean's floor.

Weather pulled back and rethrew down the ice sheet several more times until it finally conceded defeat about 590 million years ago. Defeat in the battle but not the war. It was to devise many other means of halting humanity's line of ancestry before it was too late.

With all the organisms photosynthesizing to create food, the atmosphere started to take on new characteristics. With oxygen so abundant now, our ancestors adapted to use it to grow. And grow they did. They became more complex creatures with more and more cells, each one now serving a specific function. They gifted themselves with backbones and primitive eyes. They became tetrapods, first fish with front and back fins, and then the fins grew into limbs. We don't know exactly when they did this, but they eventually used those limbs to pull themselves up onto the beaches, and they made their homes on land as amniotes. As synapsids (lizards, essentially), with new and improved lighter skulls, they flourished and became pelycosaurs, the largest and most abundant creatures on land.

But our ancestors were smarter than to just keep on growing and growing like their dinosaur cousins. While our neighbors grew to massive heights that were inarguably impressive but ultimately dangerous, our ancestors thought it best to zig while they were zagging. *When you're that big, where do you go for shelter? And how many calories will you need to maintain your girth? And what about Weather? We have a feeling that he's still planning to hit back hard. Sure, the Earth is warm and cozy now—with nary an ice cap to be seen even at the poles—but what will tomorrow bring?*

Our neighbors laughed dimly and snapped at our ancestors' tiny frames with huge, sharp teeth. Our ancestors fled to the safety of their burrows, content that they knew what they were doing.

Unassuming yet Unrelenting

By sixty million years ago, when a gigantic rock from outer space shrieked through the atmosphere and hurled itself explosively into a shallow body of water lying above sulfur deposits, our ancestors had completed their next evolutionary step into creatures that were more than prepared for such a contingency. They stayed hidden in their underground homes while a hundred-million-megaton explosion—something akin to the force of several billions of atomic bombs detonated at once—rocked the world outside. Countless creatures perished in the explosion.

Weather seized upon the opportunity, took the sulfur that had been kicked up into the sky, and rained it back down to the surface as sulfuric acid rain. It took the rest of the residue and spread it across the sky, once again blocking out the sunlight and ensconcing all living organisms in deadly cold and darkness.

Most plants, unable to photosynthesize, died, and the large dinosaurs that relied on an abundant supply of plants as a food source starved and died as well. Then the carnivorous dinosaurs that relied on the flesh of other dinosaurs to maintain the massive amounts of calories they so desperately needed succumbed to starvation and death. All told, about 70 percent of all living things on Earth hit the ground with a thud and never got up again.

And then our ancestors crawled out from their holes as furry, warm-blooded mammals, very much the same in appearance and size as the modern-day vole. They scurried around among the wreckage, climbed over the muscular leg of a *Tyrannosaurus* to get at some leaves and nuts that had been shaken loose in the commotion, and then quickly returned to their comfortable little basement apartments with an almost imperceptible extra spring in their steps.

Weather had once again tried to wipe out our lineage, but our ancestors had once again outsmarted it and used its methods against it. Had it not been for that explosion and all the hell that broke loose afterward, they quite probably would never have had the opportunity to take their next big step. But now, the biggest and most dangerous of the predators having gone up to dinosaur heaven, our ancestors were in position to take up the slack. It was finally the age of the mammal.

Now as the cool kids on the planet they used their newfound influence to partake in another huge evolutionary explosion, splintering off into countless new creatures, some of them growing to sizes that nearly rivaled their former dinosaur enemies. Some of them took over as the carnivorous bullies, feeding on the ones that thought it best to lope around and eat berries. Some even decided to return to the water, letting their noses evolve their way up onto the tops of their heads to allow for easier breathing.

And what about our ancestors? Did they take the opportunity to turn themselves into ferocious beasts with claws that dripped acid and teeth that they could shoot out of their heads as projectiles? Nope. About fifty million years ago they climbed up onto a branch and evolved into something resembling a prehistoric tree shrew and learned to mix bugs in with their regular vegetation food. While chewing on a beetle, they looked down on the carnage below, as they took the next step and evolved into the world's first primates.

Stepping Down and Stepping Out

Once again, with an ominous lightning strike, Weather reappeared ten million years later. Taking advantage of the Earth's natural progressive wobble that changes the amount of direct sunlight it gets from time to time, Weather shifted the global climate into a wintry arctic death trap and sent glaciers marching down from the poles. It continued to bring these glacial periods back from time to time. Never again with the same magnitude of earlier Snowball Earths, but with just enough intensity to shake a whole bunch of species from the family tree and into the pit of extinction below.

During these times untold numbers of still-unheard-of species vanished, their genealogical line severed where they stood. And though it came damn near close to shaking our ancestors loose once or twice, they managed to hold on barely with their newly evolved grabby prehensile fingers. Right around this time every primate on Earth—except for those in Africa and southeastern Asia—succumbed to the cold and fell away. But our ancestors had guessed correctly that it would be good to stay someplace warm. And eastern Africa, with its flat lands and lush rain forests, was just the place to call home.

That glacial period, however, was just the prelude to Weather's newest scheme. By this time the continents had floated around the Earth's magma and drifted into roughly the position we know them to assume today. However, the Indian subcontinent had yet to connect with Asia. Yet. It was just getting ready to, and twenty million years ago it slammed into lower Asia with such force that it pushed what is currently Tibet three miles higher in elevation and set off explosive volcanoes in eastern Africa. India continued pushing northward, trying in vain to get to the cooler temperatures of Russia. In doing so, it tore apart bits of eastern Africa, which resulted in the creation of the Red Sea and the Gulf of Aden around Arabia and, most important, the East African Rift, splintering from Jordan in Asia to Mozambique in eastern Africa. Not to mention the several-mile-high mountain range that accompanies it.

Our ancestors held on tight and looked out suspiciously toward India. *What fresh hell is this?* they thought. But for a long while nothing really happened. So they went about their happy primate lives. Little did they know it, but they were playing directly into Weather's hands, making themselves increasingly dependent upon the easily sought bounty of food the rain forest had to offer.

What they didn't realize was happening was that the newly formed plateau in Tibet was heating to greater intensity at its higher elevation, using that heat to feed more frequent and powerful storms in the sky above it. The water for those storms had to come from somewhere, so with a smile Weather siphoned off eastern Africa's supply and sent it over to Tibet. And meanwhile the mountains of the East African Rift began blocking the flow of moist air from the Indian Ocean into eastern Africa. Gone was the water needed for the region's rain forests to stay green and bountiful. Its fruit withered, and its trees darkened to dry husks.

Having lost their food sources, our ancestors seemed to have only slight chances of survival. It appeared, briefly, as though Weather had forced them into a checkmate. So they reached weak and trembling malnourished hands up toward Weather and proceeded to extend little prehensile middle fingers skyward. Instead of starving, they evolved their brains to a size large enough that they could think their way out of this newest predicament. With their new big brains, they surmised that to stay in a dying forest is tantamount to suicide, so a little more than three million years ago, they left the safety of the trees with their dwindling supplies of soft fruits and went off in search of new types of food on the African savannas.

To travel such distances, especially in the now-bitter Weather's streaking sunlight, they'd also need to teach themselves to walk on just two limbs. That way they could move faster while using less energy than they'd need for traveling on four. Plus they would have less sunlight pounding upon their backs, so they could stay cooler. As it turned out, our ancestors—now *Australopithecus africanus*—quite liked this new means of perambulation; it made them feel rather jaunty. And so they never went back.

Weather made certain that the going was not easy for our ancestors. It kept their lands as dry and barren as possible, so that food was scarce and demand for it high. This meant many animals all vying for the same few morsels.

Some of our ancestors' cousins, such as *Paranthropus boisei*, responded to this by growing out their jaws and molars, while developing strong chewing muscles, so that they could eat foods—nuts, tubers, and roots—that would require more grinding to be digested. Our ancestors, *Homo habilis,* chose instead to take a stab at flesh. However, being not particularly large, strong, or fast, they were in no position to chase down other animals to kill themselves. So they became scavengers, feasting on the remnants of whatever dead animal they found along their path, usually something killed by one of the big cats that roamed their neighborhood. It might seem degrading, but it was incredibly wise. That meat they ate was practically bursting with calories that their bodies could use to continue building up those big brains of theirs. Larger brains meant better problem-solving skills.

Then eventually there was more meat to be found inside the leftover bones, but they had no means of retrieving it. Unless . . .

What if they took two rocks and smashed one against the other in just the right way so that one of them ended up with a sharp, jagged side? With a thing like that, a *Homo habilis* could certainly break through hard bones. Once the bones were open, they could eat marrow to their hearts' content, ingesting more and more calories and continuing the growth of their brains.

Over the next several million years or so, they continued to get smarter, more cunning. And they'd need to be in the harsh environment that Weather had devised for them 1.8 million years ago. That was when Weather had whipped up another glacial period that drew in more great quantities of moisture from the atmosphere to feed the huge walls that crept slowly down North America and northern Europe. Less atmospheric moisture meant drier conditions for our ancestors in Africa. Finding sustenance got more difficult, so our ancestors once again needed to evolve.

In order to cover more ground and to scavenge the decreasing amount of food that was out there, our ancestors grew taller and sleeker, walking fully upright now with narrow pelvises, so as to more easily move quickly. The problem with this was that smaller hips and larger brains might seem like a huge trade up, but it had decidedly problematic effects for keeping the gene pool flowing. There was no easy way to get that big-brained baby's big head through his mom's tiny birth canal. She'd either have to go her whole life childless or risk dying during childbirth. But rather than give up and cut genealogical ties to the future, they tried something a little wacky.

Homo ergaster babies began to be born with underdeveloped brains that would grow larger in time. But that meant the baby would be unable to care for itself for several years after birth, which meant the mother would need to spend valuable time that could be spent scavenging just to keep it alive. However, leaving a baby alone to walk twenty miles in search of a dead antelope was a good way to keep the local lion population in fresh baby meat. Staying with the baby, though, meant that both mother and child would starve. So the family unit was born. Mother, father, and child would remain together until the child was old enough to go off and get some other pretty *ergaster* girl pregnant. The mother and father would need to share responsibilities.

From Cave People to Village People

In time they grew tired of scavenging. Why not just get the meat at its source? By five hundred thousand years ago, our ancestors were already killing and preparing animals on their own with surprisingly sophisticated tools, weighted throwing spears for hunting and hand axes for slicing up the kill. Then also, by this point our ancestors *Homo heidelbergensis* had moved out of their eastern African homeland into Europe and Asia, seeking new sources of food. And by two hundred thousand years ago, they'd developed language the better to brag about their own hunting skills and taunt others who slipped and fell in dung during the hunt.

These people were *Homo sapiens,* as human as you and I, 30 percent more human than sports radio callers, who have still yet to completely master the art of speech. In fact in this time lived "Eve." She's the last person along the genetic line to share a direct lineage with every person who has lived since, the great-great-super-great-grandmother of every person on Earth. And her genetic line came ridiculously close to getting wiped out just another hundred thousand years later.

We don't know exactly what Weather did a hundred thousand years ago, but whatever it was, it was totally effed up. The Earth's human population dropped down to just ten thousand people. That's about the number of people who presently live in the city of Fortuna, California. (Never heard of it? There's a reason.)

Scientists suspect that Weather had figured out some way to create a superdrought that made the survival of life in Africa, where Eve lived, next to impossible for our ancestors. And it wasn't going to get any better. Although Africa did bounce back somewhat over the next several thousand years, evidence points to a volcanic eruption in Sumatra seventy thousand years ago that sent the world into a terribly destructive ice age, with temperatures in Africa dropping by as much as 16°F, which brought more unbelievable droughts.

Given such a situation, along with the hit its population took just thirty thousand years earlier, humanity seemed doomed for sure. How ironic that the human genealogical line made it billions of years just to get snuffed out in the infancy of genuine humanity and so close to the invention of the iPhone. If they could just hold on a little bit longer. But how?

Turns out their solution was something simple, yet no other animal on Earth had tried it. They invented materialism.

Not the iPhone per se, but the iPhones of their time. They made beads out of whatever pretty things they came across—ostrich eggshells, seashells, etc.—and strung them together to make a shabby chic sort of jewelry. And then they gave them away as gifts. It sounds ridiculous, but this helped foster friendship and trading between people outside the immediate family/tribe. Communities grew larger, and people pulled together to survive the terribly cold conditions. From there villages sprang up. And then towns and cities and metropolises and countries. And we became the people we are today.

Despite Weather's best attempts, since practically the beginning of the planet, to destroy us, we survived. And not only did we survive, but we became powerful. While we once lived reactively, grasping the promise of a future with both arms in order not to find ourselves swept away by the strong currents of history, we are now proactive. We shape the planet to suit our needs. *We* build; *we* cultivate; *we* destroy; *we* push species into extinction at our whim. And practically every one of the tools that we have to do all this was given to us unintentionally by Weather and its unrelenting hatred for us.

And now, finally, we have the means of striking back.

The Vole People Strike Back

The average global temperature is all over the place when you look at it throughout the history of the planet. It goes from as high as 450°F when the Earth was just a little baby 4.5 billion years ago to sub-Antarctic lows that fell below –120°F during Weather's Snowball scheme. Over time ice ages rolled in, sucking up all available moisture into their slothlike glaciers and exposing low-lying landmasses that have not been seen since. And they were subsequently replaced by worldwide tropical temperatures that melted the polar ice caps and raised the average sea level some 213 feet higher than today. In short, there's no definitive proper global average.

We, however, have been extremely lucky to have raised our civilizations in a period of somewhat in-betweenness. This really isn't what the world is *supposed* to feel like. The last ice age that occurred 1.8 million years ago and supposedly ended around 11,000 years ago most likely never ended at all. If it had truly ended, New York and Los Angeles would be underwater. Now that might not seem like such a terrible tragedy, but it is where most of the best TV shows are filmed, so take that into consideration.

Eleven thousand years may seem like a long time for anomalous pleasant conditions, but if you take into account just how long the Earth has been here, it's nothing. Think of the life of the planet as being your own life. You just spent more than eleven thousand years reading this sentence. Eventually, if we wait around long enough, the ice will be back. And then it will go. And then it will come back. And then . . . Well, you get the idea. Each time the ice comes, expect to wait about a hundred thousand years for it to go away again. The ice won't go away for good (or at least until the next ice age) until this has happened about fifty times. And then all the ice will melt and the oceans will come rushing in. So be careful what you wish for.

So there's no point in trying to establish how hot or cold the world *should* be, but we can take a look at what we *want* it to be. Practically the entirety of human civilization has taken place during this one period of in-betweenness, so we've become really accustomed to it. In fact we've become so accustomed to it that we feel justified in despising it. *Ugh, one hundred five degrees is way too hot to go out for a walk. Let's just stay home and watch* The Simpsons. . . . *It's seven degrees outside? Screw that! I'm calling out of work.* I'm not calling all of us wussies, but our ancestors might have.

To deal with these ungodly temperatures, we did something that no other creature on Earth has ever done. We adjusted our environment to suit our own desires for comfort. If the air is uncomfortably warm and humid in a Southeast Asian jungle, what is a Bengal tiger to do? Not much; its options are somewhat limited. It can roar loudly to no one in particular, laze in the shade on a tree branch, or try to eat a *National Geographic* cameraman. None of these will serve as much more than distraction from unpleasantness, and they certainly won't do anything to improve conditions.

A human being in Wyoming, on the other hand, will walk outside, decide that the air is particularly disgusting, and then turn right around, walk back inside, and crank up the air conditioner/dehumidifier. He's then free to sit comfortably while on the couch, eating lime-flavored tortilla chips and watching a news report about a missing *National Geographic* cameraman, all in his own self-contained cool and mild indoor climate.

Think about that. When things got too hot and uncomfortable for us, we actually climbed down out of the trees and, through sheer force of will, evolved ourselves smart enough to go out and buy space heaters and desk fans.

But the really great thing about us is that we're not content to stop there. We're just now realizing that solely through our own actions, we have the power to change the world.

Humanity Makes a Difference

Over the last ten thousand years, the average global temperature has shifted up and down just a hair, with the variance never changing by more than 3.6°F. That doesn't seem like much of a variance, and it might not be over the course of millions of years. But here in this period of in-betweenness, it means a lot to us.

In the past hundred years, the average global temperature has risen by about 1°F, with the 1990s being the warmest decade of the twentieth century and 1998 the warmest year in a millennium. You might think that's a coincidence, but let me assure you that it is not. That's all us, baby, all us! We did that. By burning up the long-dead remains of dinosaurs in our cars on our way to pilates class down the street and cutting down forests to print quasihumorous books about how Weather works, we're slowly but surely raising the planet's temperature.

Practically anything we do in this superindustrialized society involves the consumption of some type of carbon-based fuel. Doesn't matter whether you're charging your laptop, turning on a black light in your dorm room, mixing a frozen margarita in the blender, hopping on the bus to avoid having to walk with your talkative neighbor to the subway stop, flying home for Thanksgiving even though you know your mother is going to ask you—yet again—if you're gay, or secretively smoking a cigarette in the privacy of the fort house your father built for you when you were seven and waiting for your obnoxious uncle's car (which also burns carbon-based fuel) to pull out of the driveway.

Coal, oil, gasoline, wood, tobacco, crack cocaine: They're all carbon-based; they originated their time on Earth as something that was a living thing, and all living things are built primarily from carbon. You may remember from Chapter 2 how plants pull carbon dioxide (CO_2) from the surrounding air and use its carbon (C) to build their cells while releasing oxygen (O_2) into the air. Wood, paper, and tobacco are obviously plant life, while coal, oil, and gasoline are derived from long-dead animals, which ate the plants and used their carbon to build their own cells. When you burn any of them, you're just reversing the process, triggering a chemical reaction that pulls O_2 from the air and fusing it once again with the C in the fuel. The end result is heat or energy plus CO_2. (Pretty much the same process happens when we breathe.)

What happens to the CO_2? It floats back into the air, where we hope it will get swallowed up by a hungry plant or dissolved into the ocean to be consumed by its algae. The animal and vegetable kingdoms have for a long time had a pretty symbiotic relationship that way.

However, that all began to change back in the nineteenth century during the Industrial Revolution, when factories began to burn huge quantities of coal to power factories. They ended up producing more CO_2 than all the plants in the world could possibly ingest themselves. Even the algae in the oceans, which have swallowed up more CO_2 than all the larger, prettier plants on Earth, could not handle the workload. And CO_2 started slowly to rise.

Today we burn up exponentially more carbon-based fuels than we did in the nineteenth century. Practically everything we do around the house requires electricity, which is still primarily generated by burning carbon-based fuels. Even when we do things that seemingly don't require energy like reading or writing sad poems about Mother Earth, we're using objects that *did* require electricity to be made. That pen and that book didn't just grow up out of the Earth that way.

On top of that, there're so many more people now than there ever were in the past. In 1804 the world's population reached one billion. Does that seem like a lot? Well, in 2011 it's estimated to hit seven billion. That's seven billion people turning on the bathroom light to relieve themselves in the middle of the night, seven billion people turning up the thermostat just a hair on a particularly chilly February afternoon, seven billion people opening the refrigerator and staring blankly at its contents when they're kinda sorta hungry but . . . eh. It adds up.

Of course the Earth would naturally be creating a certain amount of CO_2 all on its own, such as from volcanoes, and CO_2 levels have shifted throughout the Earth's history; but they've never risen as high as they have in the past few centuries, and volcanic activity can't account for the rise.

By studying the air quality of the little bubbles that

exist within extremely old Antarctic ice sheets, climatologists have discovered in the past eight hundred thousand years, the amount of CO_2 in the atmosphere has never risen more than 280 parts per million. Never, that is, until now. It's currently at 383 parts per million and rising.

So what? Why are 383 little molecules of CO_2 of any importance to anybody? The thing is, carbon dioxide is one of the most important greenhouse gases in the atmosphere, sucking up heat and keeping it from escaping into outer space. Without it—and other greenhouse gases, such as water vapor, methane, nitrous oxide, and ozone—the average global temperature would drop down from its current comfortable 59°F to a chilly –2°F. In fact, if you were to chart CO_2 levels in the atmosphere and place them beside a chart of average global temperature going back to when we started recording such things, the two charts would near perfectly fit together. Whenever there've been higher levels of CO_2, the temperature has risen, and whenever CO_2 has fallen off, the temperature has dropped.

Okay, we know that temperatures have been considerably higher in the past one hundred years, and we know that CO_2 levels have been higher in the last eight hundred thousand years. Furthermore, we can surmise that at least a nice-sized chunk of all that extra CO_2 can be attributed to our factories and vehicles and Starbucks coffee cups. If you were to factor all that together, what does it mean?

It means we're awesome! We did it! We effin' did it! For the first time in our species' history—no, for the first time in our species' lineage's history—we've got the upper hand on Weather. Us! Those little furry volelike things that went scurrying away to their underground dwellings when Weather rained down sulfuric acid on them have grown into creatures capable changing their environment on a global level. In your face, Weather! In . . . your . . . face!

The Little Chemical Compound That Could Kill Us All

To be fair, our championship work affecting the world's temperature is not our only achievement in this battle against Weather. Ever heard of a little global disaster called ozone depletion? That was us. We did that. It was kind of a little side project we had going for a while, and I guess it turned out pretty okay.

At the risk of bragging, I'd really like to fill you in on how that all went down. It started out with one man and a dream. The man was some scientist guy, and the dream was tearing a hole in the very sky itself. And all we needed to do this was a neat little family of chemical compounds with the sonorous name of chlorofluorocarbon.

Beginning in the 1930s, chlorofluorocarbons (or CFCs for short) were nonflammable, nonpoisonous chemicals (usually gases) made up of chlorine, fluorine, and carbon and used in all kinds of spiffy things. They went into the Freon we used to refrigerate our Coca-Cola and keep our living rooms cool for visiting guests. We used them as propellants in our aerosol cans to keep our bouffants stiff and our underarms smelling like lilacs. We used them to remove pesky mustard stains from our best seersucker jackets. And we used them in the foam rubber that went into our No. 1 Team giant novelty hands. On top of all that, they served another valuable purpose: They killed ozone molecules dead.

When they were sprayed out of a can or leaked from the back of a refrigerator, they didn't go anywhere. They were so inert a gas that practically nothing reacted with them. So they just floated around the beauty salon or across the kitchen until someone opened a window, and then they drifted off into the sky. Once they got up to the top of the troposphere, they still didn't go anywhere. Since they didn't mix with water, they never wound up inside a cloud droplet and fell back to Earth inside a raindrop. They were just content to hang out. Eventually they got bored and drifted higher up into the stratosphere and, ultimately, the ozone layer. And now this is where things get really cool.

Remember how we said that the ozone layer is primarily responsible for keeping harmful ultraviolet radiation from the sun from making it all the way down to Earth? Well, once a little CFC molecule gets up here, eventually it's going to come into contact with one of those ultraviolet rays. And then it completely flips out, like Bruce Banner turning into the Hulk when some guy in a diner pours soup on his head. *Chlorofluorocarbon smash!*

The ultraviolet ray throws out one of its chlorine atoms, which then reacts with the first ozone (O_3) molecule it encounters. It pulls one of the oxygen atoms loose and hurls it away. The poor ozone atom, which has no idea why it's being brutalized in such a manner, finds itself relegated to the status of oxygen (O_2). Then the chlorine atom sees another ozone molecule and does it all over again. And it keeps doing it, maybe a hundred thousand times or so, until it finally gets tired and combines with some other molecule.

Meanwhile you have all these stray oxygen atoms floating around, confused and scared, so they cling on to whatever they can find to cling on to. Mostly they end up clinging to another stray oxygen atom and settling down as a simple oxygen molecule. So, now you have two fewer O_3 molecules and three more O_2 molecules. Oxygen molecules are fine and all for breathing and junk, but we don't really have a lot of people up in the stratosphere needing to breathe. What we need is ozone to stop the ultraviolet rays before they give us cancer.

When you consider the sheer amount of CFCs that we set loose in the atmosphere and the amount of damage just one can do on its own, it's not hard to imagine why we have a great big hole in the ozone layer above Antarctica. It's pretty impressive actually. Ingenious even. And we did that just because we *can*. Simply as a warning shot across Weather's bow to let it know that we're taking control. *You see this, Weather? We're going to reach up into your sky and rip a swath of it down with our hands. We don't care if we give ourselves tumors all over our faces in the meantime. We're crazy like that.*

Ultimately we decided that giving ourselves cancer was probably not in our collective best interests, so a few decades ago we stopped using CFCs and started replacing them with other chemical compounds that aren't quite as dangerous. But there's still a lot of leftover CFCs in the sky, and there will be for at least another century or so. Consider them a reminder of the lengths to which we're prepared to go.

Finally . . . Weather, in the Palm of Our Hand

You may hear a lot of conflicting opinions about this whole global warming thing. First of all, you've got the people who say that it's not happening at all, despite the fact that the overwhelming consensus among scientists, former vice presidents, and movie stars that it is in fact happening. You can say what you want about scientists and former vice presidents—they're mostly just a bunch of geeks and dweebs—but movie stars . . . they're richer, more attractive, better dressed, and more charismatic than we are. They're our betters. If we can't trust our movie stars, who can we trust?

Then you have a bunch of people who say that yes, global warming is real, but it's not caused by humans. It's part of the natural shifting of temperatures that happens every couple thousand years, they say. To those people, I say: Are you insane? What is wrong with you? We humans just pulled off the greatest feat in the history of the world—actually changing the world—and you want to go give the credit to Weather? The same Weather that's been trying to eradicate us for several billion years. I'm sorry, but you're wrong. It was us. We did it. We did it! Even the Intergovernmental Panel on Climate Change—a bunch of meteorological and climatological experts (unfortunately, not movie stars) who were hired by the World Meteorological Organization and the United Nations to spend all their time meticulously going over charts and graphs—say that we did it. According to them, there is a greater than 90 percent chance that it was man-made, primarily caused by the burning of fossil fuels. So don't tell me that I've been keeping that tire fire going in my backyard for three years for nothing.

Of course there's a whole other group of people who think that yes, global warming is real, and yes, it is caused by humans, *but* we should try to reverse it. That same Intergovernmental Panel on Climate Change falls into this camp. My question to these people is: Why? Why should we suddenly turn around and play nice with Weather? What, you can't handle a few dozen more hurricanes every year? Are you afraid that your hair is going to get messed up when the sea levels rise and the ocean water comes pouring in through your windows? Are you afraid of droughts sweeping across the land and killing off millions and millions of people or the average global temperature rising by 8°F by the year 2100? Go to the store and buy a case of bottled water and a fan.

Finally, you have the group of people who think that yes, global warming is real, and yes, it is caused by humans, but we should embrace it. No, we should expand upon it. This group includes me and seven other people, mostly guys who hang out at the bar around the corner from my house. All I'm saying is that after billions of years we finally have Weather on the run. Why let it slip through our fingers when we could just as easily crush it in the palms of our hand? All that terrible stuff that's going to happen to us—the hurricanes and flooding and droughts and heat waves—that's just Weather acting up because it knows it's losing. If we show weakness now, who knows what Weather will come back with next century? Let's push this one to the limit.

In fact, if we play our cards right, there's actually a school of thought that says that we might be able to achieve something very interesting with all this global warming. There's a chance—not a big chance, but just a chance—that we might be able to kill Weather. But we'll have to be strong.

One way that we could do this is to push global warming to the point that we create a **runaway greenhouse effect.** What we'll do is continue to throw CO_2 into the atmosphere and therefore continue to increase its temperature. As it gets hotter, more water will get evaporated from our oceans, lakes, and rivers. And since water vapor is also a greenhouse gas, that will contribute to the rise in temperature. More warming, more evaporating, and therefore more warming. If we could keep this cycle going, we might be able to raise the temperature high enough so that all the water on Earth gets sucked off the surface and the world becomes a gigantic desert. What's Weather going to do without water, huh?

Granted, there are flaws with this plan. For one, is the water just going to float around in the sky forever? It seems likely that it'll eventually just rain back down and refill the oceans, lakes, and rivers. Plus this has never ever happened once in 4.54 billion years. We're still working this one out.

Another possibility is something called the **runaway ice age,** and this *has* happened before, but it was initiated by Weather. Imagine this: As the world warms, more and more water evaporates into the sky and becomes clouds. Lots of clouds. So many clouds that they cover the sky and block out the sun. With less solar radiation, the world will quickly drop in temperature, causing ice to form at the poles and move downward to the lower latitudes. Just like with Weather's Snowball plan, more ice means more sunlight reflected away from the Earth. And then more snow and ice. And it gets colder and colder and colder until glaciers close in over the equator, like the enmeshing cilia of the (probably by this point extinct) Venus flytrap. Clouds vanish; snow stops falling; winds die down. We will have put Weather itself into frozen hibernation.

And that's awesome.

AFTERWORD
AFTERWARDS

So, here we are. After several billion years, countless lives lost, and an unfathomable number of suede jackets ruined, we are finally in a position not only in which we can understand the unholy mechanisms of Weather and in which we can predict the devious maneuvers of Weather but in which we can *do something about it.* (I italicized that last bit because it was *extremely profound.*)

But, you may be asking, if we did *do something about it,* wouldn't we most likely be dooming ourselves to extinction along with Weather? Wouldn't this be nothing more than kamikaze tactics? Won't we all just end up dead at the end of all this?

Maybe. Very likely. Probably. But not necessarily.

You see, Weather, through constantly forcing us to find new and more difficult means of survival, may have inadvertently granted us the tools we need to save ourselves after we make life on this planet uninhabitable and generally unpleasant. Science. Our newfound understanding of Weather's DNA and our hard-won grasp of technology.

Once we've set the chain of events leading to Weather's demise in motion, we could just opt to take off. Actually. Climb inside a spaceship and take off. Earth isn't the only planet in the world, you know. We've got seven others and I don't know how many moons to choose from in this solar system alone. Surely one of them could be used as an alternative location on which to propagate our species. Let's just be a little creative.

For example, there's this theoretical thing call **terraforming** we could try. That's when you tinker with a planet's atmosphere, temperature, and ecology to make its conditions more hospitable to human life. It's sort of like gentrifying a neighborhood by opening up Starbucks and overpriced boutiques there, except you're doing it with a planet. (It's equally obnoxious and possibly unethical, but that never stopped us before.) Very likely, we'd have to import a lot of chemicals—we'd need water, oxygen, carbon dioxide, and nitrogen at the very least—to whichever planet or moon we decided upon. Mars seems like a possible choice. Or possibly Saturn's moon Titan. Or Jupiter's moons Ganymede and Europa. All of them already have atmospheres that are roughly similar to Earth's, and it's speculated that bacterial life may already have existed there, if not now, then at sometime in their pasts.

Don't get me wrong. None of this is a foregone conclusion. Still very theoretical, but it's not beyond possibility one day. We just have to ask ourselves: Are we prepared to continue putting up with Weather's unrelenting nonsense forever, or are we prepared to take charge of our destinies?

I, for one, believe that our choice is clear. I believe that we can rise above Weather and destroy it utterly. I believe that we can build a new home on a distant planet.

I believe that freed from the shackles of Weather's tyranny, we can create for ourselves a utopian world for our grandchildren. And our great-great-grandchildren. And our great-great-great-great-great-great-great-great-grandchildren. And our great-great-great-great-great-great-great-great-great-great-great-great-great-great-great-great-great-great-great-grandchildren. And our—well, you get the idea. I believe that we can finally and unequivocally be happy.

. . . Until five billion years from now, anyway, when the sun becomes a red giant and expands to several times its size before shrinking once again into an insignificant planetary nebula, thus destroying everything and everyone and every last trace of humanity.

But that's, like, our descendants' problem.

ACKNOWLEDGMENTS

The following is a severely abridged list of people who I sincerely hope never get sucked up into a tornado, struck on the head with an ice-encrusted fish, or suffer any of the other traumas that Weather has in its dread repertoire:

Susan DiClaudio, Dennis DiClaudio, Sr., Denelle DiClaudio, Anthony DiClaudio, Diandra DiClaudio, Carmen Panarello, Anthony DiMaggio, Ray DiClaudio, Rob Reinfeld, and my entire family . . .

. . . Branda Maholtz, David Cashion, Alicia Bothwell, Jennifer Tait, and everybody at Penguin Books, for their hard work and understanding . . .

. . . Emily Levin, for her empathy and patience . . .

. . . Suzanne Lanza, for her patience and empathy . . .

. . . Julie Ann Pietrangelo, Karen Lurie, Dustin Chinn, David Nagler, Eric March, Matthew Tobey, Lindsay Robertson, Michael Kraskin, Rachel Maceiras, Mary Phillips-Sandy, Wayne Gladstone, Jarrett Brilliant,

ACKNOWLEDGMENTS

Mike Alexander Mozer, and everybody at Comedy Central Digital Media . . .

. . . The Cabal, Country Bush, Jonah, everybody at the Magnet Theater, and everybody in the Philadelphia improv community . . .

. . . Paula Balzer, for her incomparable sagacity . . .

. . . and every one of my friends in Philadelphia, New York, South Jersey, and places elsewhere who have helped me to write in a bazillion different ways, both large and small, obvious and obscure, sober and not so sober. I would love to list them here in their entirety, but that task seems more daunting than trying to explain the global patterns of the wind. And I'd probably do an even worse job of it.

SOURCES AND FURTHER READING

The following books were among my most used and enjoyed sources of information during both my researching and writing processes. While *Man vs. Weather* can I hope serve as a decent enough primer for those interested in meteorology, it is in no way complete. (Lots of information had to be abandoned to make room for my idiotic ramblings.) For those interested in going deeper into the subject, the following list of books should prove helpful. They are, each one, highly recommended.

Extreme Weather by Christopher C. Burt
Freaks of the Storm by Randy Cerveny
Meteorology Today by C. Donald Ahrens
An Ocean of Air and *Snowball Earth* by Gabrielle
    Walker
The Rough Guide to Weather by Robert Henson
A Short History of Nearly Everything by Bill Bryson
Weather for Dummies by John D. Cox
Weather Tracker by Leslie Alan Horvitz
The Weather Wizard's Cloud Book by Louis D.
    Rubin, Sr. and Jim Duncan